W9-ABK-476

Rose Anita McDonnell

Catherine D. LeTourneau

Anne Veronica Burrows

Anne Brigid Gallagher

Francis H. Murphy

M. Winifred Kelly

*with*

Dr. Elinor R. Ford

LOURDES LIBRARY
CURRICULUM COLLECTION
DISCARD

Lourdes Library
Gwynedd-Mercy College
DISCARD
P. O. Box 901
Gwynedd Valley, PA 19437-0901

**Sadlier-Oxford**
A Division of William H. Sadlier, Inc.

# Table of Contents


| Textbook Chapter | | Workbook Page |
|---|---|---|
| | Title Page | i |
| | Table of Contents | ii–iv |
| | Problem-Solving Strategy: Guess and Test | 1 |
| | Problem-Solving Strategy: Hidden Information | 2 |
| | Problem-Solving Strategy: Two-Step Problem | 3 |
| | Problem-Solving Strategy: Write a Number Sequence | 4 |
| **1** | **NUMERATION, ADDITION, AND SUBTRACTION** | |
| | Expanded Form | 5 |
| | Rounding to Greatest Place | 6 |
| | Properties of Addition | 7 |
| | Estimating Sums and Differences | 8 |
| | Adding Larger Numbers | 9 |
| | Zeros in Subtraction | 10 |
| | Inverse Operations | 11 |
| | Problem-Solving Strategy: Missing/Extra Information | 12 |
| **2** | **MULTIPLICATION AND DIVISION** | |
| | Properties of Multiplication | 13 |
| | Special Patterns | 14 |
| | Estimating and Finding Products | 15 |
| | Zeros in Multiplication | 16 |
| | Exponents | 17 |
| | Short Division and Divisibility | 18 |
| | Estimating Quotients | 19 |
| | Zeros in Division | 20 |
| | Finding Quotients | 21 |
| | Order of Operations | 22 |
| | Problem-Solving Strategy: Interpret the Remainder | 23 |
| **3** | **DECIMALS: ADDITION AND SUBTRACTION** | |
| | Decimals | 24 |
| | Decimals and Expanded Form | 25 |
| | Rounding Decimals | 26 |
| | Compare and Order Decimals | 27 |
| | Estimating Decimal Sums and Differences | 28 |
| | More Adding Decimals | 29 |
| | More Subtracting Decimals | 30 |
| | Problem-Solving Strategy: Use Simpler Numbers | 31 |
| **4** | **DECIMALS: MULTIPLICATION AND DIVISION** | |
| | Multiplying Decimals by 10,100, and 1000 | 32 |
| | Estimating Decimal Products | 33 |
| | Multiplying Decimals by Whole Numbers | 34 |
| | Multiplying Decimals by Decimals | 35 |
| | More Multiplying Decimals | 36 |
| | Dividing Decimals by 10, 100, and 1000 | 37 |
| | Patterning with Tenths, Hundredths, Thousandths | 38 |
| | Estimating Quotients | 39 |
| | Dividing Decimals by Whole Numbers | 40 |
| | Dividing by a Decimal | 41 |
| | Decimal Divisors | 42 |
| | Zeros in Division | 43 |
| | Rounding Quotients | 44 |
| | Working with Decimals | 45 |
| | Scientific Notation | 46 |
| | Problem-Solving Strategy: Multi-Step Problem | 47 |
| **5** | **NUMBER THEORY AND FRACTIONS** | |
| | Fractions | 48 |
| | Finding Equivalent Fractions | 49 |
| | Prime and Composite Numbers | 50 |
| | Prime Factorization | 51 |
| | Greatest Common Factor | 52 |
| | Fractions in Simplest Form | 53 |
| | Mixed Numbers and Improper Fractions | 54 |
| | Fraction Sense | 55 |
| | Least Common Multiple | 56 |



Copyright © William H. Sadlier, Inc. All rights reserved.

Textbook Chapter | Workbook Page

Comparing Fractions ..... 57
Ordering Fractions ..... 58
Fractions, Mixed Numbers, Decimals ..... 59
Fractions: Renaming as Decimals ..... 60
Decimals as Fractions and Mixed Numbers ..... 61
Terminating and Repeating Decimals ..... 62
Problem-Solving Strategy: Find a Pattern ..... 63

## 6 FRACTIONS: ADDITION AND SUBTRACTION

Addition Properties: Fractions ..... 64
Estimating Sums and Differences ..... 65
Adding Fractions ..... 66
Adding Mixed Numbers ..... 67
Subtracting Fractions ..... 68
Subtracting Mixed Numbers ..... 69
Properties and Mixed Numbers ..... 70
Problem-Solving Strategy: Working Backwards ..... 71

## 7 FRACTIONS: MULTIPLICATION AND DIVISION

Multiplying Fractions by Fractions ..... 72
Multiplying Fractions and Whole Numbers ..... 73
Properties and the Reciprocal ..... 74
Multiplying Mixed Numbers ..... 75
Meaning of Division ..... 76
Dividing Fractions by Fractions ..... 77
Estimation in Division ..... 78
Dividing Whole Numbers ..... 79
Dividing a Mixed Number ..... 80
Order of Operations Using Fractions ..... 81
Fractions with Money ..... 82
Problem-Solving Strategy: Use a Diagram ..... 83
Problem Solving: Review of Strategies ..... 84

Textbook Chapter | Workbook Page

## 8 STATISTICS AND PROBABILITY

Graphing Sense ..... 85
Surveys ..... 86
Collecting Data ..... 87
Range, Mean, Median, and Mode ..... 88
Stem-and-Leaf Plot ..... 89
Working with Graphs and Statistics ..... 90
Making Line Graphs ..... 91
Analyzing Line Graphs ..... 92
Double Line and Double Bar Graphs ..... 93
Interpreting Circle Graphs ..... 94
Probability ..... 95
Compound Events ..... 96
Predictions ..... 97
Misleading Graphs and Statistics ..... 98
Problem-Solving Strategy: Organized List ..... 99

## 9 GEOMETRY

Congruent Segments and Angles ..... 100
Constructing Perpendicular Lines ..... 101
Measuring and Drawing Angles ..... 102
Classifying Angles ..... 103
Constructions with Angles ..... 104
Polygons ..... 105
Classifying Triangles ..... 106
Classifying Quadrilaterals ..... 107
Circles ..... 108
Classifying Space Figures ..... 109
Congruent and Similar Polygons ..... 110
Transformations ..... 111
Tessellations ..... 112
Problem-Solving Strategy: Logic/Analogies ..... 113

## 10 MEASUREMENT

Measuring Metric Length ..... 114
Measuring Metric Capacity and Mass ..... 115
Renaming Metric Units ..... 116
Relating Metric Units ..... 117


Copyright © William H. Sadlier, Inc. All rights reserved.

| Textbook Chapter | Workbook Page |
|---|---|
| Measuring Customary Length, Capacity, and Weight | 118 |
| Computing Customary Units | 119 |
| Using Perimeter | 120 |
| Area of Rectangles and Squares | 121 |
| Discovering Perimeter and Area | 122 |
| Area of Triangles and Parallelograms | 123 |
| Surface Area | 124 |
| Circumference | 125 |
| Area of a Circle | 126 |
| Volume of a Prism | 127 |
| Computing with Time | 128 |
| Problem-Solving Strategy: Use Drawings/Formulas | 129 |


## 11 RATIO, PROPORTION, AND PERCENT


| Textbook Chapter | Workbook Page |
|---|---|
| Ratio | 130 |
| Equal Ratios | 131 |
| Rates | 132 |
| Proportions | 133 |
| Solving Proportions | 134 |
| Writing Proportions | 135 |
| Using Proportions | 136 |
| Scale Drawings and Maps | 137 |
| Percent as Ratio | 138 |
| Relating Percents to Fractions | 139 |
| Relating Percents to Decimals | 140 |
| Decimals, Fractions, and Percents | 141 |
| Percents Greater Than 100% | 142 |
| Problem-Solving Strategy: Combining Strategies | 143 |


## 12 PERCENT APPLICATIONS


| Textbook Chapter | Workbook Page |
|---|---|
| Mental Math: Percent | 144 |
| Percent Sense | 145 |
| Finding a Percent of a Number | 146 |
| Missing Percent | 147 |
| Using Percent to Solve Problems | 148 |
| Finding Discount and Sale Price | 149 |
| Finding Sales Tax and Total Cost | 150 |
| Better Buy | 151 |
| Finding Commission | 152 |
| Making Circle Graphs | 153 |
| Problem-Solving Strategy: Write an Equation | 154 |


## 13 INTEGERS AND COORDINATE GRAPHING


| Textbook Chapter | Workbook Page |
|---|---|
| Integers | 155 |
| Comparing and Ordering Integers | 156 |
| Addition Model for Integers | 157 |
| Another Addition Model | 158 |
| Adding Integers | 159 |
| Subtracting Integers | 160 |
| Temperature | 161 |
| Ordered Pairs of Numbers | 162 |
| Graphing Ordered Pairs of Integers | 163 |
| Graphing Transformations | 164 |
| Problem-Solving Strategy: Make a Table | 165 |
| Problem Solving: Review of Strategies | 166 |


## 14 MOVING ON: ALGEBRA


| Textbook Chapter | Workbook Page |
|---|---|
| Algebraic Expressions | 167 |
| Equations | 168 |
| Solving Equations: Guess and Test | 169 |
| Solving Equations: Add and Subtract | 170 |
| Solving Equations: Multiply and Divide | 171 |
| Evaluating Formulas | 172 |
| Evaluating Volume Formulas | 173 |
| Multiplying Integers | 174 |
| Dividing Integers | 175 |
| Equations with Integers | 176 |
| Function Tables | 177 |
| Rational Numbers: Number Line | 178 |
| Compare/Order Rational Numbers | 179 |
| Problem-Solving Strategy: More Than One Solution | 180 |




Copyright © William H. Sadlier, Inc. All rights reserved.

# Problem-Solving Strategy: Guess and Test

Name ______________________

Date ______________________

Elena opens a book and looks at the two page numbers to which the book is opened. She says to Ed, "The product of the page numbers is 1332. What are the page numbers." What numbers should Ed name?

Guess two consecutive numbers.
Find the product to test your guess.

| First Page | 20 | 30 | 34 | 40 |
|---|---|---|---|---|
| Second Page | 21 | 31 | 35 | 41 |
| Product | 420 | 930 | 1190 | 1640 |
| Test | too low | too low | too low | too high |

So the page numbers are between 34 and 40.
Try 36 and 37.
$36 \times 37 = 1332$ Ed should name 36 and 37.

**Solve. Do your work on a separate piece of paper.**

1. The product of three consecutive numbers is 4080. What are the numbers?

______________________

2. The sum of three consecutive odd numbers is 369. What are the numbers?

______________________

3. Armando's father is 25 years older than Armando. The sum of their ages is 39. How old is Armando?

______________________

4. Angela's mother is 3 times as old as Angela. If the sum of their ages is 48, how old is Angela?

______________________

5. Diane worked twice as long as Dominic. Together they worked 12 hours. How long did Diane work?

______________________

6. Felicia has $7 more than Frank. Together they have $63. How much money does each person have?

______________________

7. Write the numbers 13, 14, 21, 35, and 42 inside the circles so that the sum along each diagonal is the same.

8. Tom was paid $4.50 per hour for painting a fence and $6.50 per hour for mowing grass. He worked 7 hours and earned $39.50. How many hours did it take him to paint the fence?

______________________

Use with text page 32.

Copyright © William H. Sadlier, Inc. All rights reserved.

# Problem Solving-Strategy: Hidden Information

Name ____________________

Date ____________________

Palani's birthday is April 24. His cousin Malia has a birthday 15 days later. On what date is Malia's birthday?

Find the hidden information needed to solve the problem.

There are 30 days in April.

Count the days from April 24 to April 30. ⟶ 6 days in April

Count on the days in May for a total of 15 days. ⟶ 9 days in May

Malia's birthday is on May 9.

**Solve. Do your work on a separate sheet of paper.**

1. Vinnie's vacation begins on July 22. It ends 14 days later. On what date does Vinnie's vacation end?

____________________

2. Rita has 4 dozen roses. If she puts 6 roses in each bouquet, how many bouquets can Rita make?

____________________

3. Shari left for work at 7:55 A.M. Her brother Eric left 15 minutes later. What time did Eric leave for work?

____________________

4. Lauri left on a vacation cruise on August 25. She returned on September 16. How many days was she on the cruise?

____________________

5. Each member of Scout Troop 164 cooked 2 eggs for breakfast. If there are 18 scouts in Troop 164, how many dozen eggs were cooked for breakfast?

____________________

6. Patty bought 24 yards of fencing to put around her garden. She only used 20 yards. How many feet of fencing did she have left over?

____________________

7. San had 6 ft of ribbon. He used 8 in. on each hat that he made. He made 7 hats. How many inches of ribbon did he have left over?

____________________

8. Herb spent 28 days exploring mountains in Colorado. How many weeks did he spend exploring?

____________________

9. Greg plans to barbecue turkey burgers for some friends. He will make 12 burgers. He wants each burger to weigh 8 ounces. How many pounds of ground turkey should he buy to make the burgers?

____________________

10. Tisha works 5 days a week as a paramedic. She has 10 holidays off and 3 weeks vacation every year. If Tisha never misses a regular day of work, how many days does she work in each year that is not a leap year?

____________________



**Use with text page 33.**

Copyright © William H. Sadlier, Inc. All rights reserved.

# Problem-Solving Strategy: Two-Step Problem

Name ______________________

Date ______________________

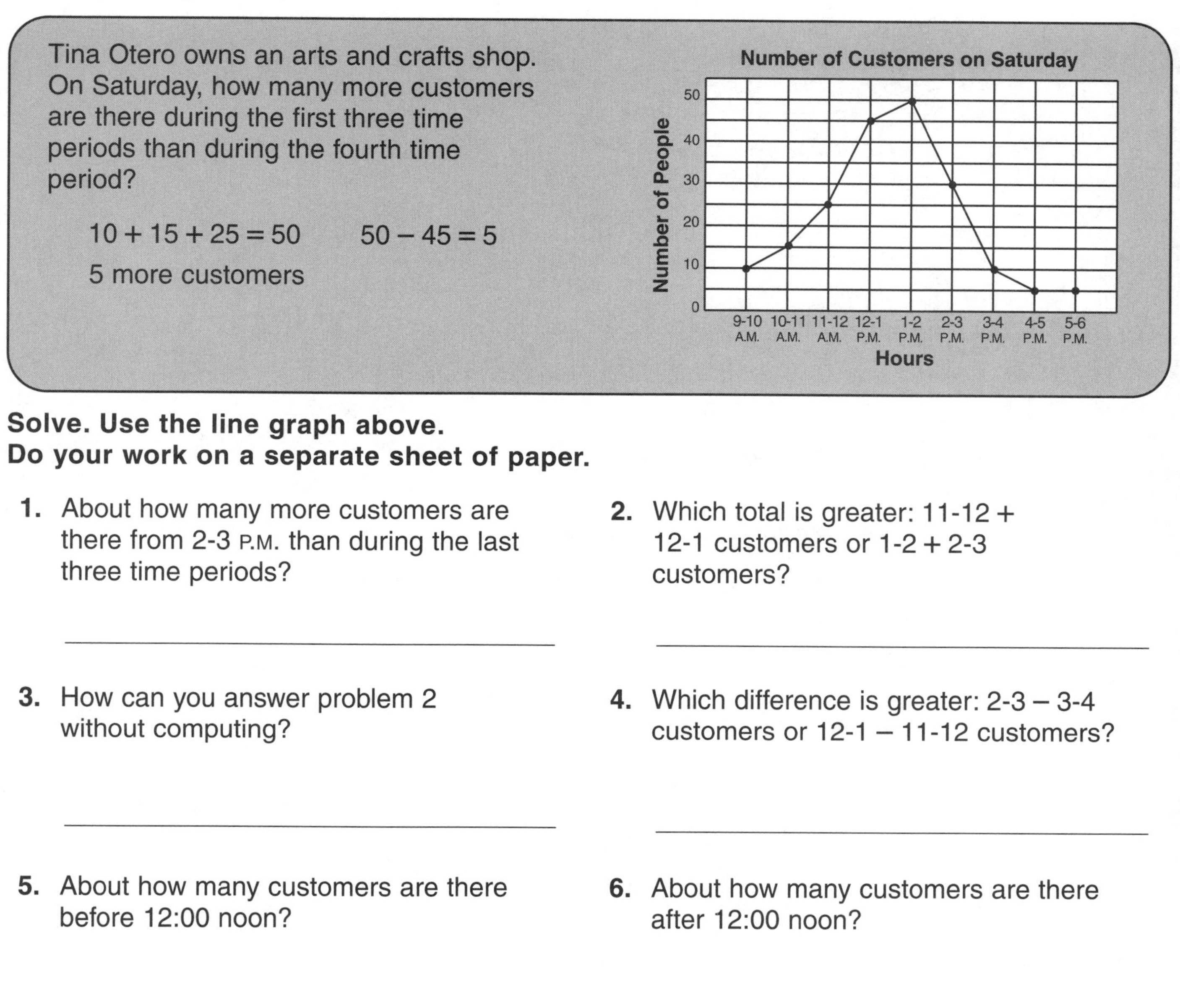


Tina Otero owns an arts and crafts shop. On Saturday, how many more customers are there during the first three time periods than during the fourth time period?

$10 + 15 + 25 = 50$ $\quad$ $50 - 45 = 5$

5 more customers

**Solve. Use the line graph above.**
**Do your work on a separate sheet of paper.**

1. About how many more customers are there from 2-3 P.M. than during the last three time periods?

______________________

2. Which total is greater: 11-12 + 12-1 customers or 1-2 + 2-3 customers?

______________________

3. How can you answer problem 2 without computing?

______________________

4. Which difference is greater: 2-3 – 3-4 customers or 12-1 – 11-12 customers?

______________________

5. About how many customers are there before 12:00 noon?

______________________

6. About how many customers are there after 12:00 noon?

______________________

7. At which times are there more than 20 customers?

______________________

8. At which times are there fewer than 10 customers?

______________________

9. Tina decides to hire a student to work from 11 A.M. to 2 P.M. Do you think this is a good decision? Why or why not?

______________________

10. If Tina were to hire extra help for any 4-hour period, which hours would they choose? Why?

______________________

Use with text page 34.

Copyright © William H. Sadlier, Inc. All rights reserved.

# Problem-Solving Strategy: Write a Number Sentence

Name ______________________

Date ______________________

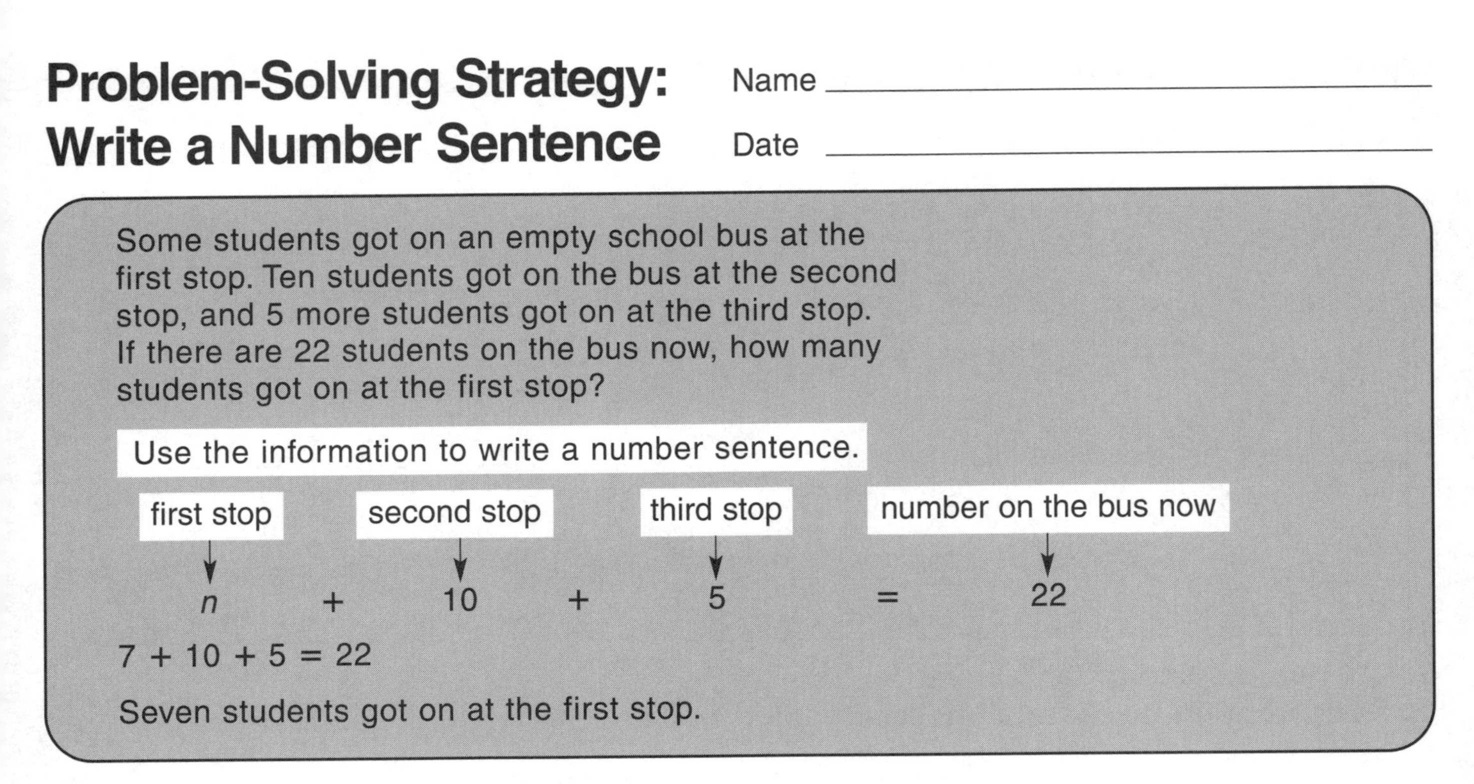

Some students got on an empty school bus at the first stop. Ten students got on the bus at the second stop, and 5 more students got on at the third stop. If there are 22 students on the bus now, how many students got on at the first stop?

Use the information to write a number sentence.



$n + 10 + 5 = 22$

$7 + 10 + 5 = 22$

Seven students got on at the first stop.

**Solve. Do your work on a separate sheet of paper.**

1. Alex weighs 12 pounds more than his younger brother. If his brother weighs 72 pounds, how much does Alex weigh?

   ______________________

2. Maria is 3 inches taller than she was this time last year. If she is 48 inches tall now, how tall was Maria this time last year?

   ______________________

3. Antwan has 12 more baseball trading cards than George. If Antwan has 28 cards, how many cards does George have?

   ______________________

4. Juan bought three books for $20.85. If each book cost the same amount, what was the cost of one book?

   ______________________

5. It takes 2 hours to complete one bus route. If the driver of the bus drives 6 hours every day, how many times does this driver cover the bus route?

   ______________________

6. The price of milk increased 8 cents a gallon this week. If the price was $2.44 per gallon last week, what is the price this week?

   ______________________

7. Uncle Darby is 12 years older than his brother. If Uncle Darby is 30 years old, how old is his brother?

   ______________________

8. If Aunt Agatha were 23 years younger, she would be 25 years old. How old is Aunt Agatha now?

   ______________________


**Use with text page 35.**
Copyright © William H. Sadlier, Inc. All rights reserved.

# Expanded Form

Name ____________________

Date ____________________

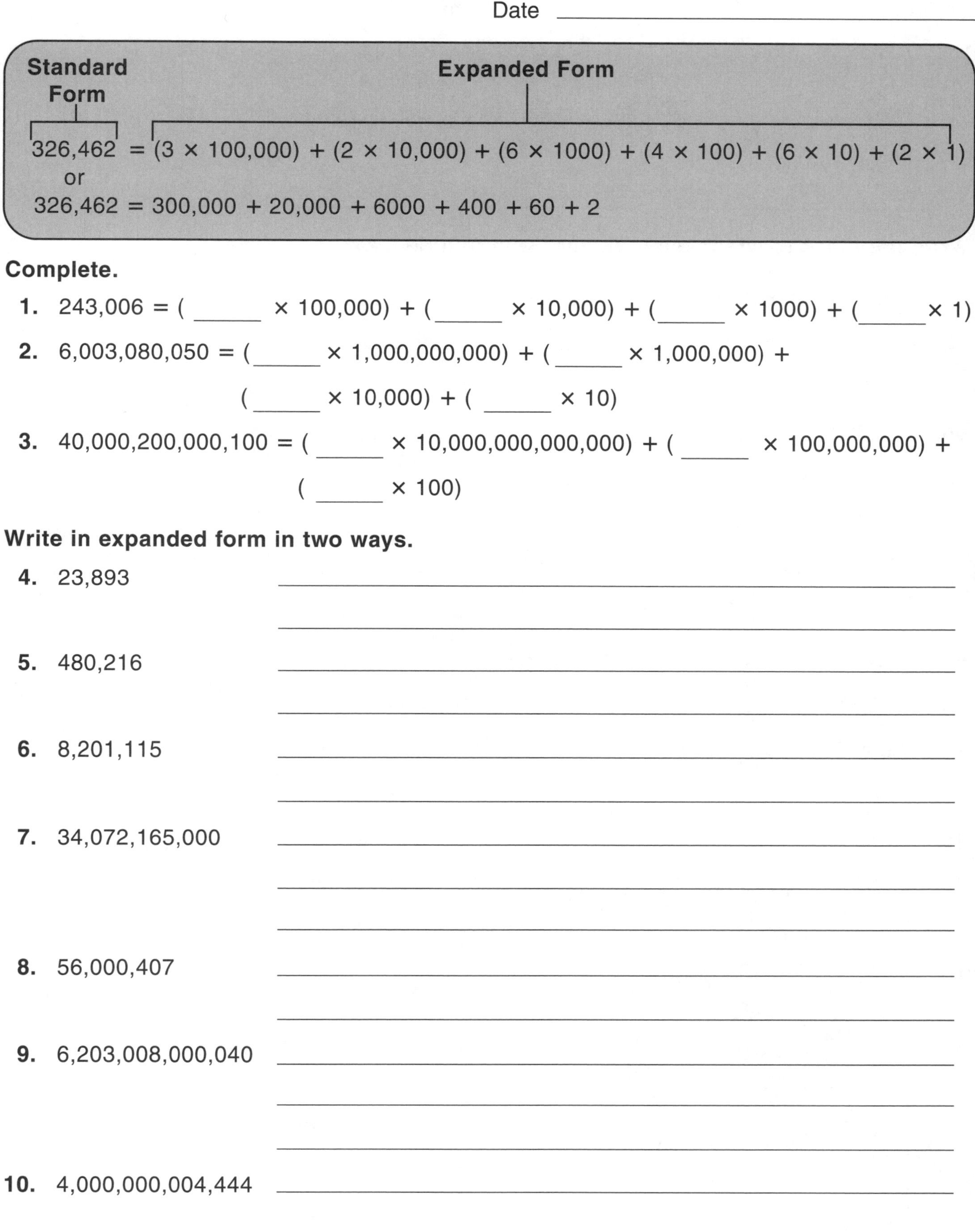

**Standard Form** | **Expanded Form**

326,462 = (3 × 100,000) + (2 × 10,000) + (6 × 1000) + (4 × 100) + (6 × 10) + (2 × 1)

or

326,462 = 300,000 + 20,000 + 6000 + 400 + 60 + 2

**Complete.**

1. 243,006 = ( _____ × 100,000) + ( _____ × 10,000) + ( _____ × 1000) + ( _____ × 1)
2. 6,003,080,050 = ( _____ × 1,000,000,000) + ( _____ × 1,000,000) + ( _____ × 10,000) + ( _____ × 10)
3. 40,000,200,000,100 = ( _____ × 10,000,000,000,000) + ( _____ × 100,000,000) + ( _____ × 100)

**Write in expanded form in two ways.**

4. 23,893 ____________________
5. 480,216 ____________________
6. 8,201,115 ____________________
7. 34,072,165,000 ____________________
8. 56,000,407 ____________________
9. 6,203,008,000,040 ____________________
10. 4,000,000,004,444 ____________________


Copyright © William H. Sadlier, Inc. All rights reserved.

# Rounding to Greatest Place

Name ____________________

Date ____________________

485,271 ⟶ 500,000

7,206,394,839 ⟶ 7,000,000,000

157,607,493 ⟶ 200,000,000

**Write the place to which each number was rounded.**

1. 17,392 to 20,000 ____________________
2. 5,436,059 to 5,000,000 ____________________
3. 257,989,164 to 300,000,000 ____________________
4. 695,733 to 700,000 ____________________
5. 82,477,016 to 80,000,000 ____________________
6. 4,565 to 5,000 ____________________
7. 7,717,912 to 8,000,000 ____________________
8. 639,726,953 to 600,000,000 ____________________

**Round each number to the greatest place.**

9. 87,002 ____________
10. 289,344 ____________
11. 7,089,233 ____________
12. 27,583,900 ____________
13. 93,804 ____________
14. 351,785 ____________
15. 5,266,801 ____________
16. 26,721,000 ____________
17. 8,504 ____________
18. 915,117,368 ____________

**Write the numbers in the table from least to greatest.**

19.

| City | Population in 1900 |
|---|---|
| Atlanta, GA | 89,872 |
| Chicago, IL | 1,698,575 |
| Detroit, MI | 285,704 |
| Los Angeles, CA | 102,479 |
| Philadelphia, PA | 1,293,697 |
| Tucson, AZ | 7,531 |

____________________

____________________

____________________

____________________



Use with Lesson 1-3, text pages 42–43.

Copyright © William H. Sadlier, Inc. All rights reserved.

# Properties of Addition

Name ____________________

Date ____________________

| Commutative Property | Associative Property | Identity Property |
|---|---|---|
| $6 + 5 = 5 + 6$<br>$11 = 11$ | $(6 + 3) + 5 = 6 + (3 + 5)$<br>$9 + 5 = 6 + 8$<br>$14 = 14$ | $14 + 0 = 14$<br>$0 + 14 = 14$ |

**Name the property of addition used.**

1. $8 + 16 = 16 + 8$ ____________
2. $3 + 0 = 3$ ____________
3. $3 + (2 + 5) = (3 + 2) + 5$ ____________
4. $11 + 8 = 8 + 11$ ____________
5. $0 + 228 = 228$ ____________
6. $(8 + 9) + 1 = 8 + (9 + 1)$ ____________

**Complete. Name the properties of addition.**

7. $8 + 7 = 7 +$ ___

    ____________________
8. $(6 + 5) + 5 = 6 + (5 +$ ___ $)$

    ____________________
9. $153 + 0 =$ ___

    ____________________
10. $7 + (13 + 10) = (7 +$ ___ $) + 10$

    ____________________
11. $6 + 10 =$ ___ $+ 6$

    ____________________
12. $(2 + 3) + 5 =$ ___ $+ (3 + 5)$

    ____________________
13. $2 + ($ ___ $+ 8) = (2 + 5) + 8$

    ____________________
14. ___ $= 5 + 0$

    ____________________
15. ___ $+ 9 = 9 + 7$

    ____________________
16. $(9 + 5) +$ ___ $= 9 + (5 + 3)$

    ____________________
17. $4 + 8 = 8 +$ ___

    ____________________
18. $(2 + 9) + 7 = 2 + ($ ___ $+ 7)$

    ____________________
19. $126 +$ ___ $= 126$

    ____________________
20. $(15 + 7) + 22 = ($ ___ $+ 15) + 22$

    ____________________
21. $m + n = n +$ ___

    ____________________
22. $0 + a =$ ___

    ____________________
23. $(a + b) + c =$ ___ $+ (b + c)$

    ____________________
24. $a + b = b +$ ___

    ____________________
25. ___ $+ 0 = x$

    ____________________
26. $x + (y + z) = (x +$ ___ $) + z$

    ____________________

**Use with Lesson 1-4, text pages 44–45.**
Copyright © William H. Sadlier, Inc. All rights reserved.

# Estimating Sums and Differences

Name ____________________

Date ____________________

| Estimate: 28,233 + 16,021 | | Estimate: 57,054 − 5863 | |
|---|---|---|---|
| **Front-end Estimation** | **Rounding** | **Front-end Estimation** | **Rounding** |
| **2**8,233 + 16,021 (about 14,000) | 30,000 + 20,000 | **5**7,054 − 5,863 | 57,000 − 6,000 |
| about **3**0,000 | about 50,000 | about **5**0,000 | about 51,000 |
| Adjusted estimate: 30,000 + 14,000 = 44,000 | | | |

**Estimate the sum or difference in two ways.**

**1.** 235 + 793

**2.** 1678 + 923

**3.** \$689.75 + 56.24

**4.** 85,734 + 18,403

**5.** \$571.14 + 830.16

**6.** 3126 − 1874

**7.** 4695 − 2902

**8.** 27,454 − 2,868

**9.** 30,090 − 19,667

**10.** 75,086 − 49,192

**11.** \$365.01 − 12.95

**12.** \$844.93 − 7.50

**PROBLEM SOLVING Use estimation.**

**13.** This is the enrollment of students in Central High School: 9th grade, 328; 10th grade, 454; 11th grade, 352; 12th grade, 398. Estimate the number of students in Central High School. ____________________

**14.** Last year 1382 yearbooks were sold. This year 916 yearbooks were sold. About how many more yearbooks were sold last year? ____________________

**15.** Ready Realtors sold one house for \$138,425 and another smaller house for \$92,525. About how much less did the smaller house cost? ____________________



**Use with Lesson 1-5, text pages 46–47.**

Copyright © William H. Sadlier, Inc. All rights reserved.

# Adding Larger Numbers

Name ______________________

Date ______________________

Add: 4,309,673 + 932,515 + 6,764,842

**First estimate.**

| | | |
|---|---|---|
| 4,309,673 | → | 4,300,000 |
| 932,515 | → | 900,000 |
| + 6,764,842 | → | + 6,800,000 |
| | | 12,000,000 |

**Then add.**

$$\begin{array}{r} \scriptstyle 2\ \ 1\,1\,2\ \ 1\,1 \\ 4{,}309{,}673 \\ 932{,}515 \\ +\,6{,}764{,}842 \\ \hline 12{,}007{,}030 \end{array}$$

**Estimate. Then find the sum.**

1. $\begin{array}{r} \$178.36 \\ +\ \ 24.67 \\ \hline \end{array}$

2. $\begin{array}{r} 27{,}268 \\ +\ 14{,}243 \\ \hline \end{array}$

3. $\begin{array}{r} 103{,}259 \\ +\ 262{,}137 \\ \hline \end{array}$

4. $\begin{array}{r} \$739.42 \\ +\ \ 20.09 \\ \hline \end{array}$

5. $\begin{array}{r} 36{,}197 \\ 12{,}506 \\ +\ 3{,}299 \\ \hline \end{array}$

6. $\begin{array}{r} 414{,}825 \\ 278{,}175 \\ +\ 131{,}231 \\ \hline \end{array}$

7. $\begin{array}{r} 721{,}399 \\ 89{,}225 \\ +\ \ 4{,}942 \\ \hline \end{array}$

8. $\begin{array}{r} 42{,}893 \\ 1{,}264 \\ +\ 29{,}731 \\ \hline \end{array}$

9. $\begin{array}{r} 7{,}365{,}909 \\ +\ 4{,}138{,}572 \\ \hline \end{array}$

10. $\begin{array}{r} 38{,}657{,}644 \\ +\ 64{,}860{,}485 \\ \hline \end{array}$

11. $\begin{array}{r} 10{,}329{,}594 \\ +\ 8{,}584{,}814 \\ \hline \end{array}$

12. $\begin{array}{r} \$628.31 \\ +\ 832.46 \\ \hline \end{array}$

**Align and add.**

13. 1,310,341 + 214,236 + 16,925 = ______________________

14. 2,424,631 + 3,180,251 + 36,297 = ______________________

15. $125.98 + $862.77 + $435.29 = ______________________

16. $739.20 + $98.99 + $67.87 = ______________________

**PROBLEM SOLVING**

17. A post office processed 2,368,721 pieces of mail this week. Last week, it processed 531,683 more pieces of mail than it did this week. How many pieces of mail did it process last week? ______________________


Copyright © William H. Sadlier, Inc. All rights reserved.

# Zeros in Subtraction

Name ______________________

Date ______________________

Subtract: 6000 − 4523

| Estimate. | Regroup as necessary. Subtract. | Check. |
|---|---|---|
| 6000 → 6000<br>− 4523 → − 5000<br>about 1000 | 9 9<br>5 10 10 10<br>6000<br>− 4523<br>1477 | 1 1 1<br>1477<br>+ 4523<br>6000 |

**Estimate. Then find the difference.**

1. 6000 − 4871
2. 9200 − 3195
3. 4030 − 268
4. 9700 − 5656
5. $27.09 − 13.27
6. 88,000 − 29,574
7. $30.90 − 6.74
8. $50.60 − 35.84
9. 14,300 − 1,034
10. 927,600 − 516,932
11. 4,230,050 − 2,129,675
12. $400.70 − 152.87
13. 8,050,009 − 34,271
14. $1000.00 − 399.86
15. 7,040,306 − 26,828
16. 2,000,000 − 1,839,839
17. $1430.03 − 1349.04
18. 9,300,700 − 819,927

**Align the numbers and then subtract.**

19. 5,000,000 − 2,692,158 = ______________
20. $800.04 − $417.88 = ______________
21. 7000 − 5632 = ______________
22. 30,600 − 19,538 = ______________
23. 900,090 − 634,291 = ______________
24. 1,000,000 − 235,895 = ______________

**PROBLEM SOLVING**

25. The Jacobs family wants to buy a car that sells for $12,000. They have saved $9472. How much more money do they need? ______________

26. A manufacturer shipped 250,000 pairs of cross trainers. Stores sold 197,658 pairs of the cross trainers. How many pairs were *not* sold? ______________



**Use with Lesson 1-7, text pages 50–51.**

Copyright © William H. Sadlier, Inc. All rights reserved.

# Inverse Operations

Name ______________________________

Date ______________________________

Addition and subtraction are inverse operations.

| | | | |
|---|---|---|---|
| | $\underline{?} - 15 = 35$ | | $\underline{?} + \$75.50 = \$98.24$ |
| | $35 + 15 = \underline{?}$ | | $\$98.24 - \$75.50 = \underline{?}$ |
| | $35 + 15 = \mathbf{50}$ | | $\$98.24 - \$75.50 = \mathbf{\$22.74}$ |
| So | $\mathbf{50} - 15 = 35$ | So | $\mathbf{\$22.74} + \$75.50 = \$98.24$ |

**Write four related sentences for these numbers.**

**1.** 29, 54, 83

______________________

______________________

______________________

______________________

**2.** \$6.27, \$3.49, \$9.76

______________________

______________________

______________________

______________________

**3.** 532, 99, 631

______________________

______________________

______________________

______________________

**Use a related sentence to find the missing number.**

**4.** $\underline{?} - 62 = 38$ ______________________

**5.** $\underline{?} + 17 = 50$ ______________________

**6.** $27 + \underline{?} = 75$ ______________________

**7.** $198 - \underline{?} = 120$ ______________________

**8.** $48 - \underline{?} = 12$ ______________________

**9.** $36 + \underline{?} = 62$ ______________________

**10.** $12 + \underline{?} = 20$ ______________________

**11.** $\underline{?} - 15 = 7$ ______________________

**12.** $a + 25 = 42$ ______________________

**13.** $56 = y + 8$ ______________________

**14.** $x - 106 = 98$ ______________________

**15.** $224 = b + 156$ ______________________

**Write the letter of the number sentence you would use. Then use a related sentence to solve.**

| | |
|---|---|
| **a.** | $375 - \underline{?} = 250$ |
| **b.** | $375 = 150 + \underline{?}$ |

**16.** Jenna has 375 stamps. She has mounted 150 of them in an album. How many more stamps does she need to mount?

______________________ ______

**17.** Luis had 375 pumpkin seeds. He gave some to Juwon and now has 250 seeds left. How many did he give to Juwon?

______________________ ______


Copyright © William H. Sadlier, Inc. All rights reserved.

# Problem-Solving Strategy: Missing/Extra Information

Name ____________________

Date ____________________

The state with the greatest area is Alaska, with an area of 589,757 sq mi. The area of Rhode Island, the smallest state, is 1214 sq mi. How much greater is the area of Alaska than the area of Lake Superior?

You do not need to know the area of the smallest state. You need information not given in the problem: the area of Lake Superior.

Use the table.
Subtract to solve: 589,757 − 31,700 = 558,057
The area of Alaska is 558,057 square miles greater than that of Lake Superior.

| Area of the Great Lakes | |
|---|---|
| Ontario | 7550 sq mi |
| Erie | 9910 sq mi |
| Michigan | 22,300 sq mi |
| Huron | 23,000 sq mi |
| Superior | 31,700 sq mi |

**PROBLEM SOLVING Use the information above. Do your work on a separate sheet of paper.**

1. Lake Erie touches Ohio and Pennsylvania. The area of Ohio is about 31,300 square miles greater than the area of Lake Erie. What is the approximate area of Ohio?

   ____________________

2. The second smallest state is Delaware. The area of Delaware is 2057 square miles. How much greater is the area of Delaware than that of Rhode Island?

   ____________________

3. The area of California is slightly more than 7 times the area of Lake Michigan. About what is the area of California?

   ____________________

4. How much greater in area than the smallest state in the United States is the largest state?

   ____________________

5. The second largest state in the United States is Texas. The area of Texas is 267,338 square miles. This is 322,419 square miles less than the area of Alaska and 108,645 square miles more than the area of California. What is the exact area of the state of California?

   ____________________

6. Nani and her brother Kekoa live in Hawaii. Nani is 2 years older than Kekoa. Nani says that the area of Hawaii is 1386 square miles less than the area of New Jersey. The area of New Jersey is 7836 square miles. What is the area of Hawaii?

   ____________________


Copyright © William H. Sadlier, Inc. All rights reserved.

# Properties of Multiplication

Name ____________________

Date ____________________

**Commutative Property**

$$\begin{array}{r} 3 \\ \times 8 \\ \hline 24 \end{array} \qquad \begin{array}{r} 8 \\ \times 3 \\ \hline 24 \end{array}$$

**Associative Property**

$(4 \times 2) \times 6 = 4 \times (2 \times 6)$

$8 \times 6 = 4 \times 12$

$48 = 48$

**Distributive Property**

$3 \times (6 + 2) = (3 \times 6) + (3 \times 2)$

$3 \times 8 = 18 + 6$

$24 = 24$

**Identity Property**

$7 \times 1 = 7 \quad 1 \times 7 = 7$

**Zero Property**

$6 \times 0 = 0 \quad 0 \times 6 = 0$

**Name the property of multiplication used.**

1. $3 \times 4 = 4 \times 3$ ____________
2. $1 \times 10 = 10$ ____________
3. $0 \times 8 = 0$ ____________
4. $7 \times 2 = 2 \times 7$ ____________
5. $2 \times (5 \times 9) = (5 \times 9) \times 2$ ____________
6. $(8 \times 2) \times 5 = 8 \times (2 \times 5)$ ____________
7. $6 \times (3 + 2) = (6 \times 3) + (6 \times 2)$ ____________
8. $4 \times (3 \times 2) = (4 \times 3) \times 2$ ____________

**Complete. Use the properties of multiplication.**

9. $45 \times 7 =$ ______ $\times 45$
10. $0 \times 5 =$ ______
11. $(4 \times 5) \times 3 =$ ______ $\times (5 \times 3)$
12. ______ $\times 6 = 6$
13. ______ $\times 9 = 9$
14. ______ $\times 10 = 0$
15. ______ $\times (2 + 8) = (6 \times 2) + (6 \times 8)$
16. $(8 \times 9) \times 2 = 8 \times ($ ______ $\times 2)$
17. $60 \times$ ______ $= 4 \times 60$
18. $3 \times (4 + 5) = (3 \times$ ______ $) + (3 \times 5)$

**Show how to use the properties of multiplication to make these computations easier. Then compute.**

19. $5 \times 12 \times 2$ ____________
20. $20 \times (5 \times 9)$
21. $8 \times 13 \times 0$ ____________
22. $2 \times 8 \times 50$ ____________
23. $(6 \times 25) \times 4$
24. $(5 \times 7) + (5 \times 4)$
25. $5 \times 31$ ____________
26. $43 \times 3$ ____________


Copyright © William H. Sadlier, Inc. All rights reserved.

# Special Patterns

Name ______________________

Date ______________________

$1 \times 57 = 57$
$10 \times 57 = 570$
$100 \times 57 = 5700$
$1000 \times 57 = 57{,}000$

$3 \times 42 = 126$
$30 \times 42 = 1260$
$300 \times 42 = 12{,}600$
$3000 \times 42 = 126{,}000$

$26{,}000 \div 1 = 26{,}000$
$26{,}000 \div 10 = 2600$
$26{,}000 \div 100 = 260$
$26{,}000 \div 1000 = 26$

$74{,}000 \div 2 = 37{,}000$
$74{,}000 \div 20 = 3700$
$74{,}000 \div 200 = 370$
$74{,}000 \div 2000 = 37$

**Complete each pattern.**

**1.** $10 \times 68 =$ ______
$100 \times 68 =$ ______
$1000 \times 68 =$ ______

**2.** $66{,}000 \div 30 =$ ______
$66{,}000 \div 300 =$ ______
$66{,}000 \div 3000 =$ ______

**3.** $40 \times 36 =$ ______
$400 \times 36 =$ ______
$4000 \times 36 =$ ______

**4.** $48{,}000 \div 80 =$ ______
$48{,}000 \div 800 =$ ______
$48{,}000 \div 8000 =$ ______

**5.** $50 \times 25 =$ ______
$500 \times 25 =$ ______
$5000 \times 25 =$ ______

**6.** $20{,}000 \div 50 =$ ______
$20{,}000 \div 500 =$ ______
$20{,}000 \div 5000 =$ ______

**Find the product.**

**7.** $64 \times 50$

**8.** $92 \times 300$

**9.** $29 \times 40$

**10.** $46 \times 2000$

**11.** $32 \times 6000$

**12.** $56 \times 500$

**13.** $75 \times 3000$

**14.** $81 \times 80$

**15.** $14 \times 900$

**16.** $63 \times 8000$

**Find the quotient.**

**17.** $63{,}000 \div 90 =$ ______
**18.** $36{,}000 \div 600 =$ ______
**19.** $14{,}000 \div 7000 =$ ______

**20.** $18{,}000 \div 200 =$ ______
**21.** $32{,}000 \div 8000 =$ ______
**22.** $21{,}000 \div 300 =$ ______

**23.** $30{,}000 \div 5000 =$ ______
**24.** $56{,}000 \div 70 =$ ______
**25.** $20{,}000 \div 400 =$ ______

**26.** $39{,}000 \div 30 =$ ______
**27.** $82{,}000 \div 200 =$ ______
**28.** $65{,}000 \div 5000 =$ ______

**29.** $84{,}000 \div 700 =$ ______
**30.** $78{,}000 \div 60 =$ ______
**31.** $90{,}000 \div 2000 =$ ______


**Use with Lesson 2-2, text pages 66–67.**
Copyright © William H. Sadlier, Inc. All rights reserved.

# Estimating and Finding Products

Name ______________________

Date ______________________

Multiply: 572 × 6189

**First estimate.**

```
  6189  ——►   6000
×  572  ——► ×  600
        about 3,600,000
```

**Then multiply.**

```
    6189
×    572
   12378  ◄— 2 × 6189
  433230  ◄— 70 × 6189
 3094500  ◄— 500 × 6189
3,540,108
```

**Estimate. Then multiply.**

| | | | |
|---|---|---|---|
| **1.** 448 × 713 | **2.** 795 × 284 | **3.** 838 × 567 | **4.** $9.26 × 361 |
| **5.** 6245 × 964 | **6.** 1781 × 857 | **7.** 2883 × 475 | **8.** $60.98 × 167 |
| **9.** 5009 × 735 | **10.** $264.51 × 543 | **11.** $612.79 × | **12.** $595.17 × 258 |

**PROBLEM SOLVING**

**13.** Last year, Fred's Appliances sold 274 color television sets. The cost of each set, including tax, was $357.44. How much money was taken in from the sale of all the television sets?

______________________


Copyright © William H. Sadlier, Inc. All rights reserved.

# Zeros in Multiplication

Name ______________________

Date ______________________

| Multiply: 560 × 487 | Multiply: 406 × 372 |
| --- | --- |
| 487<br>× 560<br>29 220 ← 60 × 487<br>243 500 ← 500 × 487<br>272,720 | 372<br>× 406<br>2 232 ← 6 × 372<br>148 800 ← 400 × 372<br>151,032 |

**Estimate. Then multiply.**

**1.** 124 × 206

**2.** 536 × 410

**3.** 159 × 203

**4.** 203 × 509

**5.** 483 × 507

**6.** 324 × 440

**7.** 602 × 306

**8.** $9.07 × 709

**9.** $2.65 × 140

**10.** 802 × 401

**11.** 1970 × 3406

**12.** 2790 × 601

**13.** 206 × 1710 = ______________

**14.** 390 × 7405 = ______________

**PROBLEM SOLVING**

**15.** A backhoe lifted 170 tons of dirt per day for 203 days. How many tons did it lift? ______________

**16.** The Apple Fruit Market received 720 cases of apples. Each case held 147 apples. How many apples did the market receive? ______________

**17.** There were 150 rows of tomato plants with 125 plants in each row. How many tomato plants were there in all? ______________

 **Use with Lesson 2-4, text pages 70–71.** Copyright © William H. Sadlier, Inc. All rights reserved.

# Exponents

Name ____________________

Date ____________________

**Standard Numeral** — **Expanded Form**

523,407 = (5 × 100,000) + (2 × 10,000) + (3 × 1000) + (4 × 100) + (7 × 1)
= $(5 \times 10^5) + (2 \times 10^4) + (3 \times 10^3) + (4 \times 10^2) + (7 \times 1)$

**Write the standard numeral.**

1. $(6 \times 10^3) + (8 \times 10^2) + (3 \times 10^1) + (4 \times 1)$ ____________
2. $(2 \times 10^5) + (3 \times 10^4) + (6 \times 10^3) + (1 \times 10^2) + (6 \times 1)$ ____________
3. $(9 \times 10^4) + (1 \times 10^3) + (6 \times 10^2) + (7 \times 10^1) + (2 \times 1)$ ____________
4. $(5 \times 10^5) + (2 \times 10^4) + (1 \times 10^3) + (4 \times 10^2) + (8 \times 10^1)$ ____________
5. $(7 \times 10^5) + (7 \times 10^4) + (7 \times 10^2) + (7 \times 10^1) + (7 \times 1)$ ____________
6. $(8 \times 10^5) + (9 \times 10^4)$ ____________

**Write in expanded form using exponents.**

7. 534 ____________________
8. 271 ____________________
9. 7025 ____________________
10. 16,530 ____________________
11. 781,003 ____________________
12. 101,407 ____________________

**Write in exponent form.**

13. 4 × 4 × 4 ____
14. 5 × 5 × 5 × 5 × 5 ____
15. 12 × 12 ____
16. 11 × 11 × 11 ____
17. 6 × 6 × 6 × 6 ____
18. 9 × 9 × 9 × 9 ____

**Write the standard numeral.**

19. $8^2$ ________
20. $3^4$ ________
21. $14^1$ ________
22. $0^5$ ________
23. $25^2$ ________
24. $7^3$ ________


Copyright © William H. Sadlier, Inc. All rights reserved.

# Short Division and Divisibility

Name ______________________

Date ______________________

Divide: 1639 ÷ 6

$$\begin{array}{r} 2\ 7\ 3\ \text{R1} \\ 6\overline{)1\ 6^{4}3^{1}9} \end{array}$$

2 × 6 = 12
16 − 12 = 4

7 × 6 = 42
43 − 42 = 1

3 × 6 = 18
19 − 18 = 1

**Divide using short division. Use R to write remainders.**

**1.** $4\overline{)6\ 4}$ **2.** $3\overline{)\$8\ 4}$ **3.** $5\overline{)2\ 1\ 4}$ **4.** $7\overline{)5\ 2\ 1}$ **5.** $9\overline{)7\ 8\ 6}$

**6.** $7\overline{)8\ 4\ 8}$ **7.** $2\overline{)3\ 5\ 7}$ **8.** $9\overline{)6\ 0\ 8\ 9}$ **9.** $6\overline{)1\ 3\ 7\ 9}$ **10.** $3\overline{)\$7\ 0\ 4\ 1}$

**11.** 73 ÷ 4 **12.** \$51 ÷ 3 **13.** 413 ÷ 6 **14.** 152 ÷ 7 **15.** \$7296 ÷ 8

**16.** 5984 ÷ 5 **17.** \$43,659 ÷ 7 **18.** 16,379 ÷ 6

**19.** 285,031 ÷ 8 **20.** 457,286 ÷ 4 **21.** 768,671 ÷ 3

**Write the divisor. Use divisibility rules to help you.**

**22.** $\begin{array}{r} 3247 \\ \overline{)9741} \end{array}$ **23.** $\begin{array}{r} 719 \\ \overline{)6471} \end{array}$ **24.** $\begin{array}{r} 684\ \text{R2} \\ \overline{)5474} \end{array}$ **25.** $\begin{array}{r} 9\ 993 \\ \overline{)89{,}937} \end{array}$ **26.** $\begin{array}{r} 4\ 223\ \text{R6} \\ \overline{)29{,}567} \end{array}$

**PROBLEM SOLVING**

**27.** Mrs. Lee bought 6 videotape movies for \$294. Each movie cost the same. How much did each cost? ______________________

**28.** The clerk in a bookstore arranged 156 books equally on 3 shelves. How many books are there on each shelf? ______________________


Copyright © William H. Sadlier, Inc. All rights reserved.

# Estimating Quotients

Name ______________________

Date ______________________

Estimate the quotient: 28,376 ÷ 62

28,376 ÷ 62

30,000 ÷ 60 = 500 ← Estimated Quotient

So 28,376 ÷ 62 ≈ 500

**Estimate the quotient.**

1. 7764 ÷ 38 ________
2. 3523 ÷ 39 ________
3. 2402 ÷ 54 ________
4. 6138 ÷ 58 ________
5. 8943 ÷ 32 ________
6. 9402 ÷ 89 ________
7. 2214 ÷ 18 ________
8. 3952 ÷ 52 ________
9. $3697 ÷ 83 ________
10. 62,093 ÷ 63 ________
11. 47,812 ÷ 47 ________
12. $92,880 ÷ 861 ________
13. 78,229 ÷ 35 ________
14. 63,029 ÷ 23 ________
15. $123,067 ÷ 549 ________

**Circle the letter of the best estimate.**

| | | a. | b. | c. | d. |
|---|---|---|---|---|---|
| 16. | 54)2935 ≈ | a. 6 | b. 60 | c. 600 | d. 6000 |
| 17. | 19)78,281 ≈ | a. 4 | b. 40 | c. 400 | d. 4000 |
| 18. | 21)13,482 ≈ | a. 5 | b. 50 | c. 500 | d. 5000 |
| 19. | 57)240,584 ≈ | a. 4 | b. 40 | c. 400 | d. 4000 |
| 20. | 37)162,432 ≈ | a. 400 | b. 4000 | c. 40 | d. 40,000 |
| 21. | 46)438,217 ≈ | a. 8000 | b. 80 | c. 8 | d. 80,000 |
| 22. | 64)619,473 ≈ | a. 10,000 | b. 100,000 | c. 1000 | d. 100 |

**PROBLEM SOLVING**

23. A small aircraft carried 6217 passengers in 28 days. If about the same number of passengers flew each day, about how many passengers were carried in one day? ________________

24. The 22 basketball teams in Big City scored a total of 8496 points. About how many points were scored by each team? ________________

25. There are 27,842 cassettes to be boxed. Each box holds 485 cassettes. About how many boxes can be filled? ________________


Copyright © William H. Sadlier, Inc. All rights reserved.

# Zeros in Division

Name ____________________

Date ____________________

Divide: 25,578 ÷ 42

$$\begin{array}{r} 6 \\ 42\overline{)25,578} \\ -\underline{252}\downarrow \\ 37 \end{array} \longrightarrow \begin{array}{r} 60 \\ 42\overline{)25,578} \\ -\underline{252}\downarrow \\ 37 \\ -\underline{\ 0} \end{array} \qquad \begin{array}{r} 609 \\ 42\overline{)25,578} \\ -\underline{252}\downarrow \\ 37 \\ -\underline{\ 0}\downarrow \\ 378 \\ -\underline{378} \\ 0 \end{array}$$

42 > 37
Write zero in the quotient.

**Estimate. Then divide and check.**

1. $8\overline{)165}$
2. $9\overline{)818}$
3. $6\overline{)1242}$
4. $23\overline{)4711}$
5. $38\overline{)7790}$
6. $27\overline{)2863}$
7. $57\overline{)22,971}$
8. $65\overline{)49,400}$
9. $54\overline{)32,778}$
10. $79\overline{)31,849}$
11. $82\overline{)41,328}$
12. $36\overline{)180,288}$

**Find the value of *n*.**

13. $n = 57,014 \div 71$
14. $n = 17,024 \div 28$
15. $546,059 \div 53 = n$
16. $188,094 \div 47 = n$


Copyright © William H. Sadlier, Inc. All rights reserved.

# Finding Quotients

Name ____________________

Date ____________________

Divide: 874 ÷ 32

$$\begin{array}{r} 2 \\ 32\overline{)874} \\ -64 \\ \hline 23 \end{array} \qquad \begin{array}{r} 27 \text{ R}10 \\ 32\overline{)874} \\ -64\downarrow \\ \hline 234 \\ -224 \\ \hline 10 \end{array}$$

Divide: $1354.50 ÷ 315

$$\begin{array}{r} \$\quad 4.\phantom{00} \\ 315\overline{)\$1354.50} \\ -1260\phantom{.50} \\ \hline 94\phantom{.50} \end{array} \qquad \begin{array}{r} \$\quad 4.30 \\ 315\overline{)\$1354.50} \\ -1260\downarrow\phantom{0} \\ \hline 94\ 5\phantom{0} \\ -94\ 5\downarrow \\ \hline 00 \end{array}$$

**Divide and check.**

**1.** 2392 ÷ 64

**2.** 9288 ÷ 43

**3.** 2118 ÷ 72

**4.** 3581 ÷ 25

**5.** $63\overline{)\$29.61}$

**6.** $58\overline{)19{,}847}$

**7.** $92\overline{)\$289.80}$

**8.** $212\overline{)9598}$

**9.** $324\overline{)\$77.76}$

**10.** $416\overline{)12{,}896}$

**11.** $146\overline{)\$233.60}$

**12.** $723\overline{)385{,}622}$

**13.** $412\overline{)335{,}792}$


Copyright © William H. Sadlier, Inc. All rights reserved.

# Order of Operations

Name ______________________

Date ______________________

Work from left to right.

Do operations within **parentheses** first.

Then **multiply** or **divide.**

Then **add** or **subtract.**

Compute: $5 + 3 \times 8 - (12 + 8)$

$5 + 3 \times 8 - 20$

$5 + 24 - 20$

$29 - 20$

$9$

**Write which operation is to be done first. Then compute.**

**1.** $24 \div 6 - 2$ ______

**2.** $15 - 5 \times 2$ ______

**3.** $42 \div 6 + 1$ ______

**4.** $8 + 2 \times 3$ ______

**5.** $(15 \div 3) - 2$ ______

**6.** $4 \times (13 - 6)$ ______

**7.** $24 \div (3 + 9)$ ______

**8.** $7 \times (4 \div 2)$ ______

**Compute.**

**9.** $3 \times 6 - 1 + 4$

**10.** $16 \div 4 \times 3 - 6$

**11.** $12 \div 2 \div 3 + 3$

**12.** $11 - 1 \times 6 + 4$

**13.** $6 \times 5 + 10 \div 2$

**14.** $3 + 9 \div 3 + 5$

**15.** $8 + 5 \times 3 - 1$

**16.** $6 + 24 \div 8 \times 2$

**17.** $4 \times 8 - (16 \div 4) \times 2$

**18.** $(20 \div 5) \times 3 + 2 \times 4$

**19.** $18 + 6 - (48 \div 6 + 2) \times 2$

**Circle the letter of the correct mathematical expression for each. Then solve.**

**20.** Sergei is saving to buy a bicycle. He saves \$15 a week from his job. His father gives him \$5 a week. After 8 weeks how much money has Sergei saved?

**a.** $8 + \$15 \times 8 + \$5$ **b.** $8 \times (\$15 + \$5)$ **c.** $(8 \times \$15) + \$5$ ______

**21.** Leila had 60 books. She gave 8 books to friends. Then she placed the remaining books equally onto 4 shelves. How many books did she place on each shelf?

**a.** $60 \div 4 - 8$ **b.** $60 - 8 \div 4$ **c.** $(60 - 8) \div 4$ ______

 Use with Lesson 2-10, text pages 82–83. Copyright © William H. Sadlier, Inc. All rights reserved.

# Problem-Solving Strategy: Interpret the Remainder

Name ______________________

Date ______________________

The county fair committee will give a free T-shirt to every 15th person who comes through the ticket gate. If 5525 people attend the fair, how many T-shirts will the committee give out?

Divide: $5525 \div 15 = \underline{?}$

```
    368 R5
15)5525
  -45
   102
   -90
    125
   -120
      5
```

Since T-shirts are given only to every 15th person, they do not need an extra shirt for the remainder.

Check:

```
   368
 ×  15
  1840
  3680
  5520
 +   5
  5525
```

The committee will give out 368 T-shirts.

**PROBLEM SOLVING Do your work on a separate sheet of paper.**

1. The fair committee wants to decorate each ride with 40 balloons. There are 13 rides. If balloons come 24 to a package, how many packages will be needed?

   ______________________

2. The Ferris wheel has seats for 36 people. If 105 people are waiting to ride the Ferris wheel, how many times will the ride need to operate in order to accommodate all of the people?

   ______________________

3. The fair committee hopes to raise $10,000 for charity. A book of 15 ride tickets costs $12. How many books does the committee need to sell to reach their goal?

   ______________________

4. A vendor plans to sell containers of juice at the fair. She packs 72 containers to a crate. How many crates will be needed if 6000 containers of juice are to be sold?

   ______________________

5. Bleacher seating for a concert on the fairgrounds consists of 6 sections. Each section seats 144 people. If 715 people attend the concert, how many sections will be needed? If 890 people attend, how many extra chairs will be needed?

   ______________________

6. The fair committee has planned a talent show for Saturday night. They have scheduled 2 hours for all the acts, including the awarding of prizes. If 9 minutes is allowed for each act, how many acts can they have? How much time will be left for the awarding of prizes?

   ______________________


Copyright © William H. Sadlier, Inc. All rights reserved.

# Decimals

Name ______________________

Date ______________________

| hundreds | tens | ones | tenths | hundredths | thousandths | ten thousandths | hundred thousandths | millionths |
|---|---|---|---|---|---|---|---|---|
| 4 | 1 | 2. | 3 | 5 | 4 | 2 | 7 | 6 |

**Read:** four hundred twelve *and* three hundred fifty-four thousand, two hundred seventy-six millionths

**Write the place of the underlined digit. Then write its value.**

**1.** 0.0<u>6</u> ______________

**2.** 0.00<u>7</u> ______________

**3.** 0.02<u>4</u>5 ______________

**4.** 0.06298<u>3</u> ______________

**5.** 24.0913<u>5</u>7 ______________

**6.** 457.098<u>1</u>23 ______________

**7.** 54.<u>9</u>267 ______________

**8.** <u>7</u>21.98634 ______________

**Use 502.638941. Name the digit in the given place.**

**9.** tens ____

**10.** thousandths ____

**11.** hundreds ____

**12.** ones ____

**13.** hundredths ____

**14.** millionths ____

**15.** ten thousandths ____

**16.** hundred thousandths ____

**17.** tenths ____

**Write the word name for each decimal.**

**18.** 0.0003 ______________________

**19.** 0.000078 ______________________

**20.** 9.42 ______________________

**21.** 1.00258 ______________________

**22.** 13.2046 ______________________

**23.** 824.015639 ______________________

______________________

**In which decimal does the digit 8 have the greater value? By how many times?**

**24.** **a.** 4.938
**b.** 2.081 ______________

**25.** **a.** 14.80
**b.** 2.008 ______________

**26.** **a.** 0.67928
**b.** 34.8097 ______________

**27.** **a.** 89.543
**b.** 0.768 ______________

**28.** **a.** 80.0007
**b.** 70.08 ______________

**29.** **a.** 245.170368
**b.** 8.002679 ______________



**Use with Lesson 3-1, text pages 94–95.**

Copyright © William H. Sadlier, Inc. All rights reserved.

# Decimals and Expanded Form

Name ____________________

Date ____________________

| Standard Form | | Expanded Form | | | |
|---|---|---|---|---|---|
| 0.40321 | = | (4 × 0.1) | + (3 × 0.001) | + (2 × 0.0001) | + (1 × 0.00001) |
| | = | 0.4 | + 0.003 | + 0.0002 | + 0.00001 |
| 2.7530 = 2.753 | = | (2 × 1) | + (7 × 0.1) | + (5 × 0.01) | + (3 × 0.001) |

**Complete.**

1. 0.2006 = (___ × 0.1) + (___ × 0.0001)
2. 6.0003000 = (___ × 1) + (___ × 0.0001)
3. 0.0407 = (___ × 0.01) + (___ × 0.0001)
4. 0.200005 = (___ × 0.1) + (___ × 0.000001)
5. 0.080732 = (8 × _______) + (7 × _______) + (3 × _______) + (2 × _______)
6. 25.400006 = (2 × _______) + (5 × _______) + (4 × _______) + (6 × _______)

**Write in expanded form in two ways.**

7. 0.8005 ____________________

8. 0.00062 ____________________

9. 0.035278 ____________________

10. 43.070280 ____________________

11. 408.20031 ____________________

12. 60,100.005 ____________________

**Write the decimal in standard form.**

13. 5 millionths ________
14. 23 ten thousandths ________
15. 2000 and 5 hundredths ________
16. 4 and 34 hundredths ________
17. 10 and 734 millionths ________
18. 48,129 hundred thousandths ________


Copyright © William H. Sadlier, Inc. All rights reserved.

# Rounding Decimals

Name ______________________

Date ______________________

**Round to the nearest:**

| tenth | hundredth | thousandth | cent |
|---|---|---|---|
| 23.8752 | 23.8752 | 23.8752 | $23.8722 |
| 23.9 | 23.88 | 23.875 | $23.87 |

**Round each number to the underlined place.**

1. $\underline{0}.73$ ________
2. $0.\underline{2}4$ ________
3. $0.6\underline{1}7$ ________
4. $25.00\underline{6}5$ ________
5. $4\underline{3}.382$ ________
6. $4.1\underline{2}27$ ________
7. $12.\underline{8}033$ ________
8. $4.66\underline{6}6$ ________
9. $0.49\underline{9}5$ ________
10. $\underline{8}6.2216$ ________
11. $400.00\underline{9}7$ ________
12. $55.5\underline{5}02$ ________

**Round each number to the greatest nonzero place.**

13. 0.64 ________
14. 8.23 ________
15. 7.008 ________
16. 0.488 ________
17. 0.86345 ________
18. 643.0029 ________
19. 159.45 ________
20. 3205.442 ________
21. 2840.75 ________

**Round to the nearest cent.**

22. $2.399 ________
23. $26.472 ________
24. $12.091 ________
25. $.029 ________
26. $.9666 ________
27. $39.995 ________
28. $56.433 ________
29. $1.998 ________
30. $.128 ________

**Place the decimal point in each numeral so that the sentence seems reasonable. Then round the decimal to the nearest tenth or nearest cent.**

31. Kira hiked 3125 miles in one hour. ________
32. Danny bought a new car for $12750279. ________
33. The temperature of the lake water was 58795°F. ________

**PROBLEM SOLVING**

34. The measurement of 1 kilometer is equal to 0.62137 miles. Round this decimal to the nearest tenth, hundredth, and thousandth. ________
35. The distance between two cities is 325.65 km. About how far apart are they to the nearest tenth of a kilometer? ________

 **Use with Lesson 3-3, text pages 98–99.** Copyright © William H. Sadlier, Inc. All rights reserved.

# Compare and Order Decimals

Name ______________________

Date ______________________

**Compare 1.590 and 1.578.**

1.590
1.578 — 1 = 1

1.590
1.578 — 5 = 5

1.590
1.578 — 9 > 7

So 1.590 > 1.578.

**Order 0.5214, 0.5380, 0.6000, 0.5372 from greatest to least.**

6 > 5 — So 0.6000 is greatest.

5 = 5 and 2 < 3 — So 0.5214 is least.

3 = 3 and 8 > 7 — So 0.5380 is second greatest.

**In order from greatest to least the decimals are:**

0.6000, 0.5380, 0.5372, 0.5214.

**Compare. Write <, =, or >.**

1. 0.09 ______ 0.0956
2. 8.07 ______ 8.189
3. 6.8 ______ 6.0087
4. 10.06 ______ 10.6715
5. 19.08 ______ 19.462
6. 36.9 ______ 39.6
7. 0.0893 ______ 0.0891
8. 20.6 ______ 20.048
9. 10.39 ______ 10.390
10. 87.642 ______ 87.6405
11. 24.24 ______ 2.42
12. 100.1 ______ 10.1

**Write in order from greatest to least.**

13. 1.44, 1.28, 1.45, 1.70

______________________

14. 0.181, 0.38, 0.139, 0.319

______________________

15. 0.74, 0.7, 0.75, 1.07

______________________

16. 0.4935, 0.492, 0.4921, 0.4853

______________________

**Write in order from least to greatest.**

17. 3.8049, 1.9942, 3.8490, 2.3756

______________________

18. 0.3886, 0.0886, 0.8386, 0.0688

______________________

19. 0.3426, 0.34, 0.342, 0.4342

______________________

20. 8.3, 8.03, 8.301, 8.3001

______________________

**PROBLEM SOLVING**
**Use the information in the table to answer each question.**

21. Which city had the most rainfall? ______________
22. Which city had the least rainfall? ______________
23. Which city had less rainfall than Dead Eye? ______________
24. Write the rainfall for the cities in order from least to greatest. ______________________

| City | Rainfall in inches |
|---|---|
| Benson | 1.6 |
| Alpha | 2.05 |
| Dead Eye | 0.92 |
| Calhoun | 0.903 |
| Essex | 1.06 |
| Freedom | 0.96 |


Copyright © William H. Sadlier, Inc. All rights reserved.

# Estimating Decimal Sums and Differences

Name ____________________

Date ____________________

0.56 + 0.24 + 0.71 is about ?

**Front-end Estimation**

$$\begin{array}{r} 0.56 \\ 0.24 \\ +\ 0.71 \\ \hline \text{about } 1.40 \end{array}$$

**Rounding**

$$\begin{array}{r} 0.56 \rightarrow 0.6 \\ 0.24 \rightarrow 0.2 \\ +\ 0.71 \rightarrow 0.7 \\ \hline \text{about } 1.5 \end{array}$$

Both 1.4 and 1.5 are reasonable estimates of the sum.

38.82 − 13.45 is about ?

**Front-end Estimation**

$$\begin{array}{r} 38.82 \\ -\ 13.45 \\ \hline \text{about } 20.00 \end{array}$$

**Rounding**

$$\begin{array}{r} 38.82 \rightarrow 40 \\ -\ 13.45 \rightarrow 10 \\ \hline \text{about } 30 \end{array}$$

Both 20 and 30 are reasonable estimates of the difference.

**Estimate the sum or difference. Use front-end estimation.**

**1.** $\begin{array}{r} 75.73 \\ +\ 62.65 \\ \hline \end{array}$ **2.** $\begin{array}{r} 76.54 \\ -\ 32.16 \\ \hline \end{array}$ **3.** $\begin{array}{r} 8.3 \\ -\ 5.4 \\ \hline \end{array}$ **4.** $\begin{array}{r} 17.98 \\ +\ 52.01 \\ \hline \end{array}$

**5.** $\begin{array}{r} 0.82 \\ -\ 0.35 \\ \hline \end{array}$ **6.** $\begin{array}{r} 85.41 \\ -\ 26.03 \\ \hline \end{array}$ **7.** $\begin{array}{r} 0.63 \\ 0.9 \\ +\ 0.35 \\ \hline \end{array}$ **8.** $\begin{array}{r} 0.5 \\ 0.37 \\ +\ 0.42 \\ \hline \end{array}$

**Estimate the sum or difference by rounding.**

**9.** $\begin{array}{r} 6.48 \\ +\ 5.84 \\ \hline \end{array}$ **10.** $\begin{array}{r} 4.287 \\ +\ 9.503 \\ \hline \end{array}$ **11.** $\begin{array}{r} 46.08 \\ -\ 21.72 \\ \hline \end{array}$ **12.** $\begin{array}{r} 84.17 \\ -\ 31.45 \\ \hline \end{array}$

**13.** $\begin{array}{r} 58.20 \\ -\ 6.82 \\ \hline \end{array}$ **14.** $\begin{array}{r} 74.53 \\ -\ 8.04 \\ \hline \end{array}$ **15.** $\begin{array}{r} 3.74 \\ 0.57 \\ +\ 4.8 \\ \hline \end{array}$ **16.** $\begin{array}{r} 45.05 \\ 12.8 \\ +\ 6.62 \\ \hline \end{array}$

**Estimate by rounding each amount to the nearest dollar.**

**17.** $\begin{array}{r} \$24.95 \\ +\ 32.63 \\ \hline \end{array}$ **18.** $\begin{array}{r} \$65.04 \\ -\ 27.95 \\ \hline \end{array}$ **19.** $\begin{array}{r} \$299.87 \\ -\ 84.15 \\ \hline \end{array}$ **20.** $\begin{array}{r} \$825.46 \\ +\ 70.28 \\ \hline \end{array}$

## PROBLEM SOLVING

**21.** Myra hiked 5.28 km, and David hiked 3.95 km. About how much farther than David did Myra hike. ____________________

**22.** Jessie bought shirts for $12.95, $10.50, $13.52, $11.48, and $9.89. About how much was the total cost of the shirts? ____________________


Copyright © William H. Sadlier, Inc. All rights reserved.

# More Adding Decimals

Name ______________________

Date ______________________

Add: 8.35 + 0.7995 + 15

- Line up the decimal points.
- Add zeros as needed.
- Write the decimal point in the sum.

$$\begin{array}{r} {}^{1\,1\,1}\phantom{0000} \\ 8.3500 \\ 0.7995 \\ +\ 15.0000 \\ \hline 24.1495 \end{array}$$

**Estimate. Then find the sum.**

1. $\begin{array}{r} 10.47 \\ +\ 0.78 \\ \hline \end{array}$ 2. $\begin{array}{r} 2.32 \\ +\ 3.5 \\ \hline \end{array}$ 3. $\begin{array}{r} \$18.96 \\ +\ \ 23.08 \\ \hline \end{array}$ 4. $\begin{array}{r} 29.2 \\ +\ 36.59 \\ \hline \end{array}$ 5. $\begin{array}{r} 16.2 \\ +\ 8.49 \\ \hline \end{array}$

6. $\begin{array}{r} 0.47 \\ 0.5 \\ 0.78 \\ +\ 0.29 \\ \hline \end{array}$ 7. $\begin{array}{r} 1.68 \\ 3.7 \\ 6.34 \\ +\ 9.5 \\ \hline \end{array}$ 8. $\begin{array}{r} 12.09 \\ 14.04 \\ 20.35 \\ +\ \ 3.6 \\ \hline \end{array}$ 9. $\begin{array}{r} 15.04 \\ 3.12 \\ 10.02 \\ +\ 0.46 \\ \hline \end{array}$ 10. $\begin{array}{r} \$10.36 \\ 3.14 \\ 8.24 \\ +\ \ \ 2.03 \\ \hline \end{array}$

11. $\begin{array}{r} 6.412 \\ 5.93 \\ +\ 9.482 \\ \hline \end{array}$ 12. $\begin{array}{r} 19.27 \\ 25.6 \\ +\ \ 6.9 \\ \hline \end{array}$ 13. $\begin{array}{r} 415.5 \\ 8.924 \\ 13.57 \\ +\ \ \ 2.6 \\ \hline \end{array}$ 14. $\begin{array}{r} 2.90 \\ 0.48 \\ 8.754 \\ +\ 62.951 \\ \hline \end{array}$ 15. $\begin{array}{r} 3.7465 \\ 8.2001 \\ 9.3561 \\ +\ 40.0008 \\ \hline \end{array}$

**Align and estimate. Then add.**

16. 0.3 + 1.04 + 2.4 + 4.07 ________

17. 1.08 + 7.5 + 4.9 + 0.06 ________

18. 12.14 + 7.06 + 31.2 + 5.04 ________

19. 0.16 + 2 + 0.08 + 1.32 ________

20. $3.70 + $9.03 + $4.06 ________

21. 32.065 + 0.16 + 3.294 ________

**PROBLEM SOLVING**

22. A train traveled 36.2 km the first hour and 32.8 km, 39.1 km, and 40.03 km for the next three hours. How far did the train travel in the four ______________

23. Last year Mary weighed 46.05 kg. If she gained 3.5 kg since then, what is her present weight? ______________

24. Harry, Devin, and Teo ran a 3-man relay race. Harry ran 6.7 km, Devin ran 7 km, and Teo ran 8.32 km. How far did they run in all? ______________

 Copyright © William H. Sadlier, Inc. All rights reserved. 

# More Subtracting Decimals

Name ______________________

Date ______________________

Subtract: 4.7 − 1.8265

- Line up the decimal points.
- Add zeros as needed.
- Write the decimal point in the difference.

```
  16 9 9
3 17 10 10 10
4.7000
−1.8265
 2.8735
```

**Estimate. Then find the difference.**

| | | | | |
|---|---|---|---|---|
| **1.** 0.84 − 0.57 | **2.** 0.60 − 0.02 | **3.** 0.71 − 0.329 | **4.** 9.45 − 6.7 | **5.** $75.80 − 9.22 |
| **6.** 0.6932 − 0.3481 | **7.** 0.5 − 0.3889 | **8.** 0.81 − 0.687 | **9.** 10.43 − 4.921 | **10.** 72.1 − 12.385 |
| **11.** 36.2 − 24.3295 | **12.** $12.00 − 8.32 | **13.** 16.537 − 8.7 | **14.** $13.00 − 7.94 | **15.** 4.013 − 0.0987 |

**Compare. Write <, =, or >.**

**16.** 26.7814 − 2.3219 ______ 24.459

**17.** 19.1 − 4.87 ______ 14.2

**18.** 6 − 3.295 ______ 2

**19.** 1 − 0.0793 ______ 0.9206

**20.** 14.232 − 6 ______ 8.232

**21.** 26.22 − 0.0049 ______ 26.275

**Align and estimate. Then subtract.**

**22.** 0.8625 − 0.7937 = ________

**23.** 55 − 0.3277 = ________

**24.** 5.736 − 4.0004 = ________

**25.** 605.2427 − 8.2427 = ________

**26.** 10.43 − 4 = ________

**27.** 50.1 − 3.427 = ________

**PROBLEM SOLVING**

**28.** The sum of a number and 5.32 is 7.8. Find the number. ________

**29.** The length of one trail is 102.9 m. A second trail is 111.09 m. How much longer is the second trail? ________

 **Use with Lessons 3-8 and 3-9, text pages 108–111.** Copyright © William H. Sadlier, Inc. All rights reserved.

# Problem-Solving Strategy: Use Simpler Numbers

Name ______________________

Date ______________________

Kathy wants to buy a bicycle that costs $149.95. She has saved $54.35 so far. Last week she earned $28.50. This week she earned $22.75. How much money does she need to buy the bicycle?

Substitute simpler numbers.
Cost of bicycle minus the money Kathy has so far equals the money she needs.

$150 − ($50 + $30 + $20) = $50

Now solve the problem using the actual numbers.

$149.95 − ($54.35 + $28.50 + $22.75) = $44.35 Kathy needs $44.35.

**PROBLEM SOLVING Do your work on a separate sheet of paper.**

1. At the beginning of the month, Craig had $384.37 in his checking account. During the month, he wrote checks for $29.50, $16.85, $44.90, and $127.38. He made a deposit of $585.50. How much did Craig have in his checking account at the end of the month?

______________________

2. Karen and Karim are playing a game with Bob and Betty. Karen scored 24 points during her turn, and Karim scored 18 points during his turn. Bob scored 21 points. How many points must Betty score during her turn if Bob and Betty are to win the game?

______________________

3. Elly, Marge, and Bryan baby-sit to earn money. Last month, Elly earned $27.50. Marge earned $8.25 more than Elly. Bryan earned $12.40 more than Marge. How much did Bryan earn?

______________________

4. Paula, Willis, and Ted are training for a marathon. Each day Paula runs 4.75 miles. Willis runs 0.3 mile less than that, and Ted runs 1.55 miles more than Willis. How far does Ted run each day?

______________________

5. Shana's balance in her credit card account was $278.76 at the beginning of the month. During the month, she made a payment of $55.00. She also charged items that cost $19.55 and $17.60. If the interest charge for the month was $3.90, what was her account balance at the end of the month?

______________________

6. Raul and Marta are planning to hike from base camp to Doone, which is 89.7 km away. If they hike 27.5 km the first day, 33.85 km the second day, and 28.35 km the third day, will they reach Doone?

______________________


Copyright © William H. Sadlier, Inc. All rights reserved.

# Multiplying Decimals by 10, 100, and 1000

Name ______________________

Date ______________________

**Patterns:**

| | | |
|---|---|---|
| $10 \times 0.652 = 6.52$ | $10 \times 1.58 = 15.8$ | $10 \times 0.8 = 8$ |
| $100 \times 0.652 = 65.2$ | $100 \times 1.58 = 158$ | $100 \times 0.8 = 80$ |
| $1000 \times 0.652 = 652$ | $1000 \times 1.58 = 1580$ | $1000 \times 0.8 = 800$ |

**Multiply:**

| | |
|---|---|
| $10 \times 0.127 = 1.27$ | 1 zero: Move 1 place to the right. |
| $100 \times 0.039 = 3.9$ | 2 zeros: Move 2 places to the right. |
| $1000 \times 0.006 = 6$ | 3 zeros: Move 3 places to the right. |
| $1000 \times 2.400 = 2400$ | 3 zeros: Move 3 places to the right. Write 2 zeros as placeholders. |

**Find the products. Use the patterns.**

1. $10 \times 0.583 =$ ______
   $100 \times 0.583 =$ ______
   $1000 \times 0.583 =$ ______
2. $10 \times 0.002 =$ ______
   $100 \times 0.002 =$ ______
   $1000 \times 0.002 =$ ______
3. $10 \times 1.34 =$ ______
   $100 \times 1.34 =$ ______
   $1000 \times 1.34 =$ ______
4. $10 \times 0.41 =$ ______
   $100 \times 0.41 =$ ______
   $1000 \times 0.41 =$ ______
5. $10 \times 1.4 =$ ______
   $100 \times 1.4 =$ ______
   $1000 \times 1.4 =$ ______
6. $10 \times 0.5 =$ ______
   $100 \times 0.5 =$ ______
   $1000 \times 0.5 =$ ______

**Multiply.**

7. $10 \times 0.02 =$ ______
8. $100 \times 0.061 =$ ______
9. $1000 \times 0.057 =$ ______
10. $100 \times 0.5 =$ ______
11. $10 \times 0.4 =$ ______
12. $100 \times 24.8 =$ ______
13. $1000 \times 0.004 =$ ______
14. $10 \times 0.0067 =$ ______
15. $1000 \times 1.9 =$ ______

**Find the products. Then write them in order from least to greatest.**

16. **a.** $10 \times 0.65 =$ ______ **b.** $100 \times 0.064 =$ ______ **c.** $1000 \times 0.063 =$ ______

    ______________________

17. **a.** $1000 \times 0.0083 =$ ______ **b.** $100 \times 0.004 =$ ______ **c.** $10 \times 0.0073 =$ ______

    ______________________

**Write the missing factor.**

18. ______ $\times 0.107 = 1.07$
19. ______ $\times 3.09 = 309$
20. $100 \times$ ______ $= 0.4$
21. $10 \times$ ______ $= 0.003$
22. ______ $\times 0.523 = 523$
23. $1000 \times$ ______ $= 3.15$

**PROBLEM SOLVING**

24. One bottle of glucose solution contains 0.25 L. How much glucose solution is there in 10 bottles? in 100 bottles?

    ______________________

25. One kilogram is equivalent to 2.2046 pounds. To how many pounds is 10 kilograms equivalent? 100 kilograms?

    ______________________


Copyright © William H. Sadlier, Inc. All rights reserved.

# Estimating Decimal Products

Name ______________________

Date ______________________

| Estimate: 4.0452 × 22.63 | Estimate: 0.89 × 57.13 | Estimate: 3.14 × 9.75 |
|---|---|---|
| 4.0452 → 4<br>× 22.63 → × 20<br>80 | 0.89 → 0.9<br>× 57.13 → × 60<br>54 | 3.14 → 3<br>× 9.75 → × 10<br>30 |
| So 4.0452 × 22.63 ≈ 80. | So 0.89 × 57.13 ≈ 54. | So 3.14 × 9.75 ≈ 30. |
| Both factors are rounded down. The actual product is *greater than* 80. | Both factors are rounded up. The actual product is *less than* 54. | One factor is rounded down, and one factor is rounded up. The actual product is *close to* 30. |

**Estimate each product. Then tell whether the actual product is *greater than*, *less than*, or *close to* the estimated product.**

1. 4.32 × 8.1 ______
2. 8.86 × 3.5 ______
3. 5.19 × 7.8 ______
4. 2.07 × 12.43 ______
5. 18.946 × 7.005 ______
6. 15.62 × 0.85 ______
7. 33.81 × 18.009 ______
8. 47.105 × 5.547 ______

**Estimate the product.**

9. 20.703 × 3.8
10. 23.9 × 0.9
11. 4826 × 0.47
12. 414 × 0.75
13. 5.2 × 61.92 ______
14. 1.2 × 3.5 × 16.8 ______
15. 10.5 × 7.05 × 32.1 ______

**Estimate by rounding *one* factor to the nearest 10, 100, or 1000.** Use mental math.

16. 10.1 × 0.456 ______
17. 9.78 × 2.1 ______
18. 95 × 3.205 ______
19. 132 × 67.2 ______
20. 984 × 0.306 ______
21. 101.99 × 0.0057 ______

**PROBLEM SOLVING**

22. Mrs. Jones bought 17 cartons of fruit juice that each cost $2.89 a carton. About how much did she spend? ______

23. A freight train travels 53.7 miles an hour. At this rate of speed, about how far will it travel in 12.1 hours? ______


Copyright © William H. Sadlier, Inc. All rights reserved.

# Multiplying Decimals by Whole Numbers

Name ______________________

Date ______________________

| Multiply: 12 × 9.623 | Multiply: 36 × $28.41 |
| --- | --- |
| 9.623<br>× 12<br>19246<br>96230<br>115.476<br>(3 decimal places) | $28.41<br>× 36<br>17046<br>85230<br>$1022.76<br>(2 decimal places) |

**Write the decimal point in each product.**

1. 0.617 × 8 = 4 9 3 6
2. 0.6 × 21 = 1 2 6
3. 2.04 × 23 = 4 6 9 2
4. 3.09 × 36 = 1 1 1 2 4
5. 0.301 × 47 = 1 4 1 4 7

**Estimate. Then multiply.**

6. 0.5 × 34
7. 0.39 × 56
8. 0.72 × 18
9. 0.069 × 4
10. 3.002 × 43
11. $3.29 × 3
12. $43.25 × 74
13. $82.15 × 6
14. $524.20 × 8
15. $5.43 × 11

**Estimate. Then find the product.**

**16.** 8 × 0.7 = ________ **17.** 4 × $.86 = ________ **18.** 3 × 0.285 = ________

**19.** 25 × $4.31 = ________ **20.** 12 × $28.13 = ________ **21.** 87 × 0.62 = ________

**22.** 9 × 4.218 = ________ **23.** 56 × 0.65 = ________ **24.** 88 × 0.06 = ________

**PROBLEM SOLVING**

**25.** Harry bought 4 dozen eggs at $1.09 a dozen. How much did he pay for the eggs? ________________

**26.** Lacey bought 15 T-shirts for the hiking club. Each shirt cost $12.75. How much did Lacey pay for the shirts? ________________


Copyright © William H. Sadlier, Inc. All rights reserved.

# Multiplying Decimals by Decimals

Name ______________________

Date ______________________

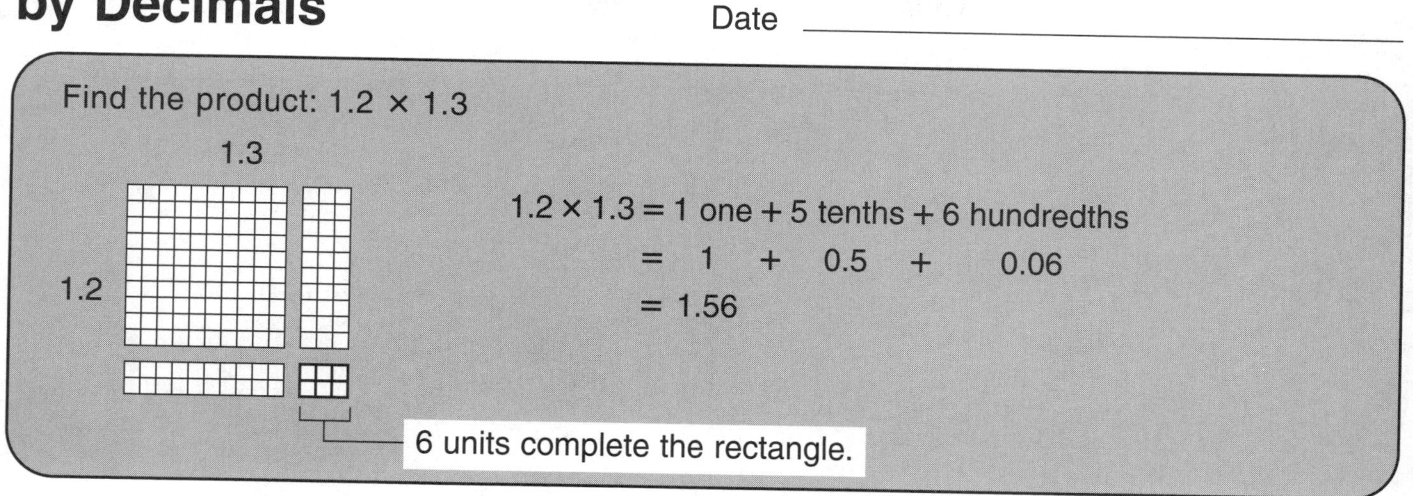


**Complete the rectangle. Find the product.**

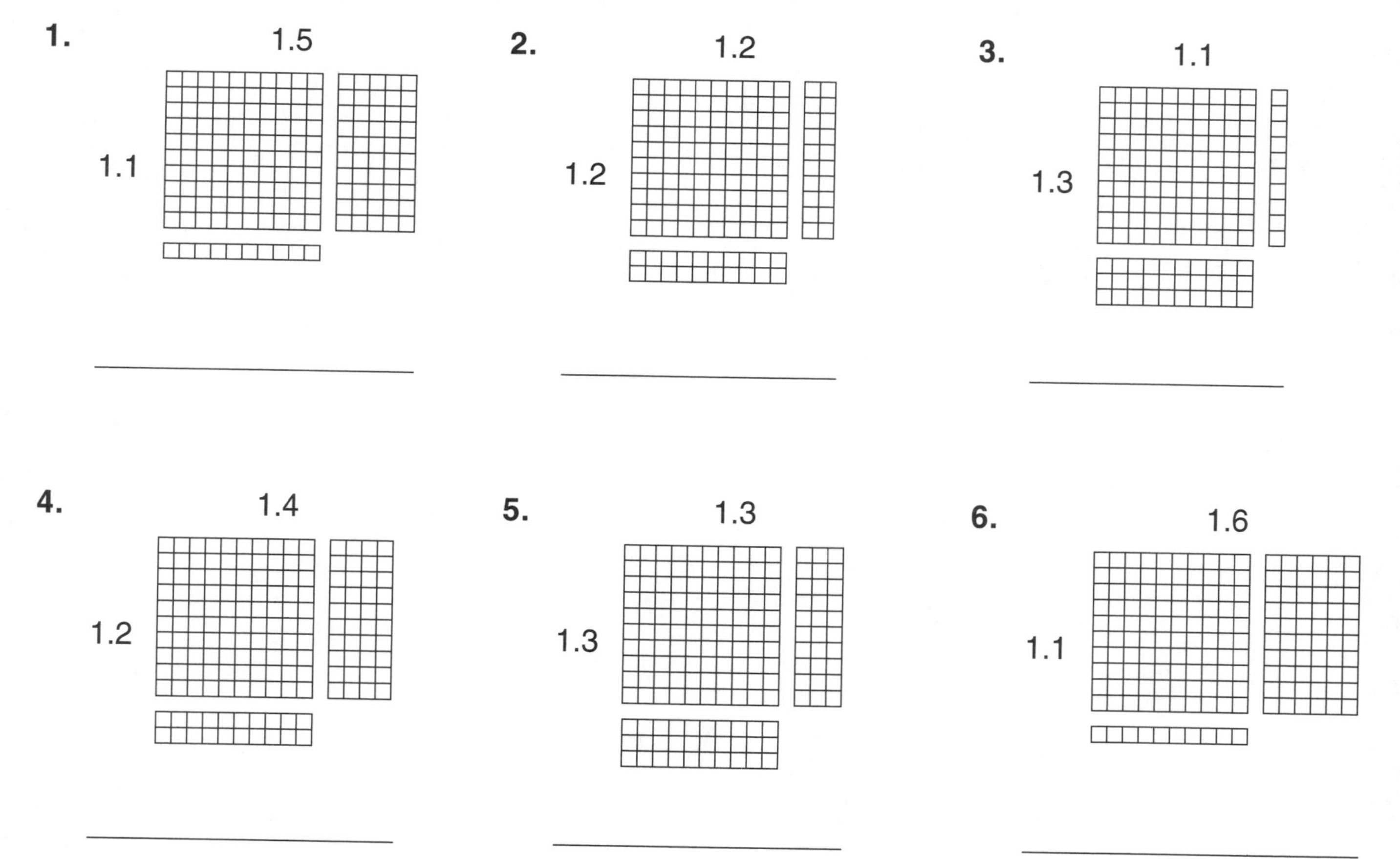


**PROBLEM SOLVING**

7. When you multiply a decimal with tenths by another decimal with tenths, how many decimal places will there be in the product? ______________________


Copyright © William H. Sadlier, Inc. All rights reserved.

# More Multiplying Decimals

Name ______________________________

Date ______________________________

Multiply: 0.04 × 1.02

```
   1.02  ← 2 decimal places
 × 0.04  ← 2 decimal places
 0.0408  ← 4 decimal places
           Write 1 zero.
```

Multiply: $1.26 × 0.03

```
  $1.26  ← 2 decimal places
 × 0.03  ← 2 decimal places
 $.0378  ← 4 decimal places
```

Round to the nearest cent.
$.0378 ⟶ $.04

**Write the decimal point in each product.** Write zeros where necessary.

1. 0.37 × 0.65 = 2405
2. 0.48 × 0.05 = 240
3. 0.419 × 0.9 = 3771
4. 62.35 × 0.26 = 162110
5. 16.79 × 1.8 = 30 222

**Estimate. Then multiply. Round to the nearest cent when necessary.**

6. 0.003 × 0.6
7. 0.715 × 0.02
8. 0.45 × 0.8
9. $ 7.02 × 1.65
10. $ .40 × 3.18

11. 14.035 × 6.8
12. 0.038 × 0.21
13. $ 156.70 × 2.4
14. 0.123 × 0.64
15. 6.030 × 5.22

16. 0.72 × 0.3 = ______
17. 0.13 × $85.28 = ______
18. 3.29 × 43.53 = ______
19. 2.03 × 1.2 × 0.5 = ______
20. 0.2 × 0.4 × 0.6 = ______
21. 0.04 × 0.1 × 0.3 = ______

**PROBLEM SOLVING**

22. Mary had 0.8 gallon of milk. She gave 0.3 of the milk to Tom. How much milk did Tom get? ______________

23. A square yard of carpeting costs $46.50. What is the cost of 3.8 square yards? ______________

24. Chain costs $2.36 per foot. What is the cost of 7.8 feet of chain? ______________



**Use with Lesson 4-5, text pages 130–131.**

Copyright © William H. Sadlier, Inc. All rights reserved.

# Dividing Decimals by 10, 100, and 1000

Name ____________________

Date ____________________

**Patterns:**

315.4 ÷ 10 = 31.54
315.4 ÷ 100 = 3.154
315.4 ÷ 1000 = 0.3154

27.5 ÷ 10 = 2.75
27.5 ÷ 100 = 0.275
27.5 ÷ 1000 = 0.0275

0.21 ÷ 10 = 0.021
0.21 ÷ 100 = 0.0021
0.21 ÷ 1000 = 0.00021

**Divide:**

5.1 ÷ 10 = 0.51 — 1 zero: Move 1 place to the left.

1 8.2 ÷100 = 0.182 — 2 zeros: Move 2 places to the left.

3 4 6 ÷1000 = 0.346 — 3 zeros: Move 3 places to the left.

0 0 7.9 ÷1000 = 0.0079 — 3 zeros: Move 3 places to the left. Write 2 zeros as placeholders.

**Complete each table. Use the patterns.**

| | Number | ÷10 | ÷100 | ÷1000 |
|---|---|---|---|---|
| **1.** | 16.2 | | | |
| **2.** | 8.9 | | | |
| **3.** | 165 | | | |
| **4.** | 0.72 | | | |

| | Number | ÷10 | ÷100 | ÷1000 |
|---|---|---|---|---|
| **5.** | 1679 | | | |
| **6.** | 56.29 | | | |
| **7.** | 0.286 | | | |
| **8.** | 126.93 | | | |

**Divide.**

**9.** 8 ÷ 10 = ________
**10.** 93 ÷ 10 = ________
**11.** 0.47 ÷ 10 = ________
**12.** 518 ÷ 10 = ________
**13.** 0.8 ÷ 100 = ________
**14.** 6 ÷ 100 = ________
**15.** 2.54 ÷ 100 = ________
**16.** 0.08 ÷ 100 = ________
**17.** 3427.1 ÷ 100 = ________
**18.** 1 ÷ 1000 = ________
**19.** 3.006 ÷ 1000 = ________
**20.** 82.45 ÷ 1000 = ________

**Write the missing number.**

**21.** 725.3 ÷ ______ = 7.253
**22.** ______ ÷ 10 = 0.014
**23.** ______ ÷ 100 = 0.076
**24.** ______ ÷ 1000 = 3.175
**25.** ______ ÷ 100 = 0.32
**26.** ______ ÷ 1000 = 0.021

**Compare. Write $<$, $=$, or $>$.**

**27.** 341 ÷ 100 _____ 34.1 ÷ 1000
**28.** 0.17 ÷ 100 _____ 0.0017 × 10

**PROBLEM SOLVING**

**29.** Driving at a speed of 100 km per hour, how long will it take to drive a distance of 475.1 km? ____________


Copyright © William H. Sadlier, Inc. All rights reserved.

# Patterning with Tenths, Hundredths, Thousandths

Name ______________________

Date ______________________

**Patterns:**

85 ÷ 0.1 = 850
85 ÷ 0.01 = 8500
85 ÷ 0.001 = 85,000

415.6 ÷ 0.1 = 4156
415.6 ÷ 0.01 = 41,560
415.6 ÷ 0.001 = 415,600

**Divide:**

5.2 4 ÷ 0.1 = 52.4 — 1 decimal place in the divisor: Move 1 place to the right.

5.2 4 ÷ 0.01 = 524 — 2 decimal places in the divisor: Move 2 places to the right.

5.2 4 0 ÷ 0.001 = 5240 — 3 decimal places in the divisor: Move 3 places to the right. Write 1 zero.

**Complete each table. Use the patterns.**

| | Number | ÷0.1 | ÷0.01 | ÷0.001 |
|---|---|---|---|---|
| 1. | 23 | | | |
| 2. | 439 | | | |
| 3. | 7.1 | | | |
| 4. | 15.3 | | | |

| | Number | ÷0.1 | ÷0.01 | ÷0.001 |
|---|---|---|---|---|
| 5. | 10 | | | |
| 6. | 2.78 | | | |
| 7. | 24.14 | | | |
| 8. | 326.89 | | | |

**Divide.**

9. 47 ÷ 0.1 = ________
10. 52 ÷ 0.01 = ________
11. 78 ÷ 0.001 = ________
12. 589 ÷ 0.1 = ________
13. 423 ÷ 0.01 = ________
14. 932 ÷ 0.001 = ________
15. 0.6 ÷ 0.1 = ________
16. 0.65 ÷ 0.01 = ________
17. 3.4 ÷ 0.001= ________
18. 253.8 ÷ 0.01 = ________
19. 3.78 ÷ 0.001= ________
20. 9 ÷ 0.1= ________

**Compare. Write <, =, or >.**

21. 7.5 ÷ 0.01 _____ 17.5 ÷ 0.001
22. 7.1 ÷ 0.01 _____ 71 ÷ 0.1
23. 96.7 ÷ 0.01 _____ 967 ÷ 0.01
24. 52 ÷ 0.01 _____ 52 ÷ 0.1

**Find the missing number.**

25. 78 ÷ _____ = 7800
26. 0.7 × _____ = 0.07
27. 3.2 ÷ _____ = 3200
28. _____ × 0.01 = 78.92
29. 42.5 × _____ = 0.0425
30. 1.94 ÷ _____ = 0.0194

**PROBLEM SOLVING**

31. How many dimes are in $45? ____________________


Copyright © William H. Sadlier, Inc. All rights reserved.

# Estimating Quotients

Name ______________________

Date ______________________

| Estimate: 65.4 ÷ 0.83 | Estimate: \$.41 ÷ 5.8 | Estimate: 14.9 ÷ 5.1 |
|---|---|---|
| Dividend > Divisor: Quotient > 1 | Dividend < Divisor: Quotient < 1 | **Front End:** 10 ÷ 5 = 2 |
| 65.4 ÷ 0.83 ↓ 64 ÷ 0.8 = 80 | \$.41 ÷ 5.8 ↓ \$.42 ÷ 6 = \$.07 | **Compatible Numbers:** 15 ÷ 5 = 3 |
| | | **Rounding:** 10 ÷ 5 = 2 |
| So 65.4 ÷ 0.83 is about 80. | So \$.41 ÷ 5.8 is about \$.07. | So 14.9 ÷ 5.1 is about 2 or 3. |

**Estimate to place the decimal point in the quotient.**

1. 45.9 ÷ 7.2 = 6 3 7 5
2. 14.28 ÷ 3.4 = 4 2
3. 55.8 ÷ 6 = 9 3
4. 8.17 ÷ 4.3 = 1 9
5. \$48.15 ÷ 5 = \$9 6 3
6. 286.44 ÷ 15.5 = 1 8 4 8

**Estimate each quotient.** Use compatible numbers.

7. 31.8 ÷ 4.1 ______
8. 12.4 ÷ 6.5 ______
9. 50.2 ÷ 6.9 ______
10. 27.3 ÷ 8.5 ______
11. 58.3 ÷ 7.9 ______
12. 35.1 ÷ 4.2 ______
13. 62.34 ÷ 8.98 ______
14. 18.3 ÷ 2.75 ______
15. 321.6 ÷ 8.2 ______
16. 146.4 ÷ 6.8 ______
17. 211.3 ÷ 19.5 ______
18. 485.7 ÷ 50.4 ______
19. 590 ÷ 32.4 ______
20. 900.67 ÷ 41.8 ______
21. 241.34 ÷ 23.72 ______

**Compare. Write <, =, or >.**

22. 5 ÷ 9 ______ 1
23. 11.8 ÷ 3.2 ______ 1
24. 1 ______ 18.3 ÷ 0.496
25. 0.87 ÷ 6.2 ______ 1
26. 3.5 ÷ 47.1 ______ 1
27. 1 ______ 0.2 ÷ 0.4
28. 1.29 ÷ 0.78 ______ 1
29. 1 ______ 1.22 ÷ 0.248
30. 2.79 ÷ 0.4 ______ 1

**Complete. Estimate each quotient.**

| | | Front End | Compatible Numbers | Rounding |
|---|---|---|---|---|
| **31.** | 44.7 ÷ 5.2 | | | |
| **32.** | 32.4 ÷ 8.4 | | | |
| **33.** | 52 ÷ 5.9 | | | |
| **34.** | 283.7 ÷ 7.2 | | | |
| **35.** | 487.9 ÷ 18.1 | | | |

**PROBLEM SOLVING**

36. Tony paid \$48.75 for 7.8 yards of fabric.
About how much per yard was the fabric? ______________________


Copyright © William H. Sadlier, Inc. All rights reserved.

# Dividing Decimals by Whole Numbers

Name ______________________________

Date ______________________________

Divide: $56.84 \div 7$

Write the decimal point of the quotient directly above the decimal point of the dividend.

$$7\overline{)56.84}$$

Divide as you would with whole numbers.

$$\begin{array}{r} 8.12 \\ 7\overline{)56.84} \\ -56\phantom{.84} \\ \hline 0\,8\phantom{4} \\ -7\phantom{4} \\ \hline 14 \\ -14 \\ \hline 0 \end{array}$$

Check.

$$\begin{array}{r} 8.12 \\ \times \quad 7 \\ \hline 56.84 \end{array}$$

**Circle the letter of the correct answer.**

**1.** $6\overline{)0.78}$ **a.** 1.3 **b.** 0.13 **c.** 13 **d.** 0.013

**2.** $84\overline{)26.88}$ **a.** 0.32 **b.** 0.032 **c.** 3.2 **d.** 32

**Divide and check.**

**3.** $9\overline{)\$28.35}$ **4.** $4\overline{)45.12}$ **5.** $2\overline{)0.286}$ **6.** $5\overline{)7.20}$

**7.** $8\overline{)\$9.76}$ **8.** $7\overline{)19.4691}$ **9.** $3\overline{)22.008}$ **10.** $4\overline{)0.852}$

**11.** $12\overline{)4.344}$ **12.** $23\overline{)9.89}$ **13.** $38\overline{)\$66.88}$ **14.** $17\overline{)\$21.76}$

**Compare. Write $<$, $=$, or $>$.**

**15.** $0.81 \div 3$ _____ $8.1 \div 3$

**16.** $0.72 \div 4$ _____ $0.72 \div 0.4$

**17.** $9\overline{)0.45}$ _____ $9\overline{)0.045}$

**18.** $6\overline{)2.4}$ _____ $0.6\overline{)0.24}$

**PROBLEM SOLVING**

**19.** Jamail paid $13.92 for 16 identical pens. How much did each pen cost? ______________

**20.** Nine packages weigh a total of 21.06 kg. Find the weight of 1 package if they all weigh the same. ______________


Copyright © William H. Sadlier, Inc. All rights reserved.

# Dividing by a Decimal

Name ____________________

Date ____________________

Divide: 23.04 ÷ 0.6 = ?

| Move the decimal points in the divisor and in the dividend. | Place the decimal point in the quotient. Divide. | Check. |
|---|---|---|
| $0.6\overline{)23.04} \longrightarrow 6\overline{)230.4}$ | $0.6\overline{)23^{5}0^{2}.4}$ with quotient 38.4 | 38.4 × 0.6 = 23.04 |

**Circle the letter of the correct answer.**

1. $0.5\overline{)3.5}$ **a.** 0.7 **b.** 7 **c.** 70 **d.** 0.07
2. $0.2\overline{)0.16}$ **a.** 0.8 **b.** 0.08 **c.** 8 **d.** 80
3. $0.7\overline{)49.7}$ **a.** 0.71 **b.** 7.1 **c.** 71 **d.** 710
4. $0.4\overline{)8.48}$ **a.** 212 **b.** 2.12 **c.** 0.212 **d.** 21.2
5. $0.3\overline{)2.4}$ **a.** 0.8 **b.** 0.08 **c.** 8 **d.** 80

**Divide and check.**

6. $0.9\overline{)8.1}$ 7. $0.6\overline{)5.4}$ 8. $0.8\overline{)3.2}$ 9. $0.2\overline{)1.4}$ 10. $0.5\overline{)4.5}$

11. $0.3\overline{)0.21}$ 12. $0.7\overline{)0.35}$ 13. $0.4\overline{)3.24}$ 14. $0.6\overline{)1.26}$ 15. $0.8\overline{)6.48}$

16. $0.2\overline{)64.4}$ 17. $0.5\overline{)85.5}$ 18. $0.8\overline{)3.36}$ 19. $0.9\overline{)5.166}$ 20. $0.4\overline{)8.52}$

21. $0.7\overline{)15.12}$ 22. $0.3\overline{)14.673}$ 23. $0.6\overline{)12.72}$ 24. $0.8\overline{)5.272}$ 25. $0.2\overline{)65.8}$

**PROBLEM SOLVING**

26. The length of a quilt is 2.4m. Each square patch is 0.2m a side. How many patches are in a row that runs the length of the quilt?

____________________

**Use with Lesson 4-10, text pages 140–141.**

Copyright © William H. Sadlier, Inc. All rights reserved.

# Decimal Divisors

Name ____________________

Date ____________________

Divide: 0.2232 ÷ 0.024

Move the decimal points in the divisor and in the dividend.

$0.024\overline{)0.2232} \longrightarrow 24\overline{)223.2}$

Divide.

$$\begin{array}{r} 9.3 \\ 24\overline{)223.2} \\ -\underline{216\phantom{.2}} \\ 82 \\ -\underline{82} \\ 0 \end{array}$$

Check.

$$\begin{array}{r} 9.3 \\ \times\ 0.024 \\ \hline 372 \\ 186\phantom{0} \\ \hline 0.2232 \end{array}$$

**Move the decimal point in the divisor and the dividend to show the quotient.**

1. $6.2\overline{)177.94}$ (quotient: 28.7)
2. $0.03\overline{)0.678}$ (quotient: 22.6)
3. $0.09\overline{)3.627}$ (quotient: 40.3)
4. $34.6\overline{)93.42}$ (quotient: 2.9)
5. $0.24\overline{)386.88}$ (quotient: 1612.)
6. $0.04\overline{)32.44}$ (quotient: 811.)

**Find the quotient.**

7. $2.3\overline{)96.6}$
8. $0.32\overline{)0.064}$
9. $5.7\overline{)1.3965}$
10. $0.35\overline{)28.875}$
11. $0.43\overline{)38.786}$
12. $6.2\overline{)386.88}$

**Divide and check.**

13. 57.312 ÷ 0.72 = _____
14. 156.51 ÷ 0.37 = _____
15. 0.056 ÷ 0.08 = _____

**PROBLEM SOLVING**

16. In a race, a car went 283.4 miles in 2.6 hours. Find the average speed per hour. ____________


Copyright © William H. Sadlier, Inc. All rights reserved.

# Zeros in Division

Name ______________________

Date ______________________

| Divide: 21 ÷ 5.25 | Divide: 0.043 ÷ 0.5 |
| --- | --- |
| $5.25\overline{)21.00} \rightarrow 525\overline{)2100}$ quotient 4; −2100; 0. Write 2 zeros as placeholders. | $0.5\overline{)0.043} \rightarrow 5\overline{)0.430}$ quotient 0.086; −40; 30; −30; 0. Write a zero. |

**Divide. When needed, write zeros as placeholders in the dividend.**

1. $0.4\overline{)0.6}$
2. $0.6\overline{)12}$
3. $7.5\overline{)9}$
4. $0.64\overline{)0.8}$
5. $0.6 \div 1.2 =$ ____
6. $3 \div 8 =$ ____
7. $0.06 \div 0.025 =$ ____
8. $0.01 \div 0.008 =$ ____

**Divide. Write zeros in the quotient as needed.**

9. $4\overline{)0.16}$
10. $3\overline{)0.21}$
11. $7\overline{)0.287}$
12. $4.1\overline{)0.123}$
13. $3.6\overline{)0.144}$
14. $4\overline{)1.208}$
15. $3.4\overline{)0.017}$
16. $5\overline{)1.505}$

**PROBLEM SOLVING**

17. Samantha can run 1.68 miles in 12 minutes. What is her average distance per minute? ____________

18. Jack bought 6.5 pounds of grapes for $7.02. What is the cost of one pound of grapes? ____________


Copyright © William H. Sadlier, Inc. All rights reserved.

# Rounding Quotients

Name ____________________

Date ____________________

| $5 \div 6 = \underline{?}$ Round to the nearest tenth. | $\$1.50 \div 8 = \underline{?}$ Round to the nearest cent. | $0.83 \div 7 = \underline{?}$ Round to the nearest thousandth. |
|---|---|---|
| $6\overline{)5.00}$ = 0.8**3**; 3 < 5 Round down. | $8\overline{)\$1.500}$ = $ .18**7**; 7 > 5 Round up. | $7\overline{)0.8300}$ = 0.118**5**; 5 = 5 Round up. |
| $5 \div 6 \approx 0.8$ | $\$1.50 \div 8 \approx \$.19$ | $0.83 \div 7 \approx 0.119$ |

**Divide. Round to the nearest tenth.**

1. $6\overline{)0.4}$
2. $13\overline{)8}$
3. $4.7\overline{)30}$
4. $3.2\overline{)12}$
5. $0.6\overline{)1.4}$
6. $3.4\overline{)0.6}$
7. $4.1\overline{)5.5}$
8. $0.7\overline{)0.9}$

**Divide. Round to the nearest hundredth or nearest cent.**

9. $9\overline{)12}$
10. $7\overline{)8}$
11. $4\overline{)7.1}$
12. $8\overline{)3.9}$
13. $4.2\overline{)6.6}$
14. $0.9\overline{)6.2}$
15. $3.4\overline{)0.6}$
16. $8.1\overline{)2.3}$
17. $0.07\overline{)1.7}$
18. $8\overline{)\$4.21}$
19. $7\overline{)\$6.33}$
20. $3\overline{)\$9.47}$

**Divide. Round to the nearest thousandth.**

21. $6\overline{)0.7}$
22. $8\overline{)4.19}$
23. $3\overline{)1.024}$
24. $0.7\overline{)4.5}$
25. $12\overline{)0.911}$
26. $7.1\overline{)3.117}$
27. $0.07\overline{)0.1}$
28. $13\overline{)141}$

 **Use with Lesson 4-13, text pages 146–147.** Copyright © William H. Sadlier, Inc. All rights reserved.

# Working with Decimals

Name ____________________

Date ____________________

| | Addition Properties | Multiplication Properties |
|---|---|---|
| Commutative | $3.4 + 0.2 = 0.2 + 3.4$ | $3.4 \times 0.2 = 0.2 \times 3.4$ |
| Associative | $0.5 + (1.7 + 0.8) = (0.5 + 1.7) + 0.8$ | $0.5 \times (1.7 \times 0.8) = (0.5 \times 1.7) \times 0.8$ |
| Identity | $6.8 + 0 = 6.8 \quad 0 + 6.8 = 6.8$ | $6.8 \times 1 = 6.8 \quad 1 \times 6.8 = 6.8$ |

Zero Property of Multiplication $0.7 \times 0 = 0 \quad 0 \times 0.7 = 0$

Distributive Property of Multiplication over Addition $4.1 \times (0.2 + 0.5) = (4.1 \times 0.2) + (4.1 \times 0.5)$

**Name the property of addition or multiplication used.**

1. $6.2 \times (1.1 + 0.3) = (6.2 \times 1.1) + (6.2 \times 0.3)$ ____________________
2. $0.5 + (1.7 + 4.2) = (0.5 + 1.7) + 4.2$ ____________________
3. $0.9 + 7.6 = 7.6 + 0.9$ __________
4. $1.7 \times 1 = 1.7$ __________
5. $0.8 \times 0 = 0$ __________
6. $8.9 + 0 = 8.9$ __________

**Complete. Use the properties of addition and multiplication.**

7. $9.06 \times 1 =$ ______
8. $0 \times 5.32 =$ ______
9. $0.1 \times 6.77 = 6.77 \times$ ______
10. $0 + 8.12 =$ ______
11. ______ $\times 1 = 0.006$
12. $0.3 + 7.21 =$ ______ $+ 0.3$
13. $(6.3 + 0.7) + 1.8 = 6.3 + ($______ $+ 1.8)$
14. $1.8 \times (0.7 + 0.4) = (1.8 \times$ ______$) + (1.8 \times 0.4)$
15. $0.15 \times (1.2 \times 0.64) = ($______ $\times 1.2) \times 0.64$
16. $n \times (1.4 + 6.2) = (n \times 1.4) + (n \times$ ______$)$

**Name the property of addition or multiplication you can use to make each computation easier. Then compute.**

17. $4 \times (2.5 \times 7.8) =$ ______ ____________________
18. $(0.4 \times 5.6) \times 25 =$ ______ ____________________
19. $(0.3 \times 9.8) \times 0 =$ ______ ____________________
20. $4 \times (2.5 + 25) =$ ______ ____________________
21. $1.6 + 2.9 + 0.4 =$ ______ ____________________
22. $7.36 \times 45.2 \times 0 =$ ______ ____________________

**Divide. Use mental math.**

23. $8 \div 0.5 =$ _____
24. $33 \div 0.5 =$ _____
25. $7 \div 0.25 =$ _____
26. $30 \div 0.25 =$ _____

 Copyright © William H. Sadlier, Inc. All rights reserved. 

# Scientific Notation

Name ____________________

Date ____________________

| Write 156,000,000 in scientific notation. | Write $4.17 \times 10^5$ in standard form. |
|---|---|
| 156,000,000 → $1.56 \times 10^8$. 1.56 is greater than 1 but less than 10. 8 places moved. The power of 10 is $10^8$. So $156{,}000{,}000 = 1.56 \times 10^8$. | $4.17 \times 10^5 = 4.17000 = 417{,}000$. Move 5 places to the right. So $4.17 \times 10^5 = 417{,}000$. |

**Write in scientific notation.**

**1.** 120,000 ________ **2.** 485,000 ________ **3.** 25,000 ________

**4.** 2,750,000 ________ **5.** 714,500 ________ **6.** 47,000,000 ________

**7.** 143,700,000 ________ **8.** 1,505,000 ________ **9.** 28,003,000 ________

**10.** 6,202,000 ________ **11.** 300,000,000 ________ **12.** 6500 ________

**Write in standard form.**

**13.** $7 \times 10^2$ ________ **14.** $8 \times 10^3$ ________ **15.** $9 \times 10^5$ ________

**16.** $2.4 \times 10^3$ ________ **17.** $1.9 \times 10^4$ ________ **18.** $3.75 \times 10^5$ ________

**19.** $2.01 \times 10^5$ ________ **20.** $4.3 \times 10^6$ ________ **21.** $7 \times 10^8$ ________

**22.** $3.436 \times 10^6$ ________ **23.** $7.005 \times 10^7$ ________ **24.** $9.415 \times 10^8$ ________

**25.** $6.012 \times 10^5$ ________ **26.** $8.02 \times 10^9$ ________ **27.** $5.2866 \times 10^5$ ________

**PROBLEM SOLVING**

**28.** Neptune is about 2,800,000,000 miles from the Sun. Write this number in scientific notation. ____________

**29.** Saturn is abut 900,000,000 miles from the Sun. Write this number in scientific notation. ____________

**30.** Jupiter is about $4.8 \times 10^8$ miles from the Sun. Write this number in standard form. ____________


Copyright © William H. Sadlier, Inc. All rights reserved.

# Problem-Solving Strategy: Multi-Step Problem

Name ____________________

Date ____________________

Ms. Hayashi buys 5 lb of chicken at \$2.19 a pound, 2 lb of fresh green beans at \$1.20 a pound, and a 64-ounce bottle of apple juice for \$1.49. How much did she spend in all?

**Step 1:** Find the cost of the chicken.
5 × \$2.19 = \$10.95

**Step 2:** Find the cost of the green beans.
2 × \$1.20 = \$2.40

**Step 3:** Add to find the total cost of the three items.
\$10.95 + \$2.40 + \$1.49 = \$14.84

Ms. Hayashi spent \$14.84 in all.

**PROBLEM SOLVING Do your work on a separate sheet of paper.**

1. Yoshi buys a T-shirt for \$15.95 and socks for \$7.95. He gives the clerk a 20-dollar bill and a 10-dollar bill. How much change should Yoshi receive?

   ____________________

2. If Bettinia drove for 3.8 hours at a constant speed of 56 mph and Willis drove for 4.1 hours at a constant speed of 52 mph, who drove farther? How much farther?

   ____________________

3. Pat wants to buy a CD player that costs \$86.25. He has saved \$60. He earns \$5.25 an hour working after school and can save half of his total earnings. How many hours will he have to work to buy the CD player?

   ____________________

4. On Saturdays Diego works 8 hours and earns \$6.50 an hour. Sabrina works 6 hours and earns \$8.95 an hour. Who earns more? How much more?

   ____________________

5. Natasha has to have \$250 for a vacation trip. She has saved \$69 and has received \$25 as a gift towards the vacation money. If she saves \$12 each week, how many weeks will it take her to reach her goal of \$250?

   ____________________

6. The price of a 6-ounce can of Kitty Kat turkey is \$.49. The price of a 4-ounce can of Cougar Cat turkey is \$.32. How much can you save by buying 12 ounces of the less expensive brand?

   ____________________

7. A nature trail is 2.4 km in length, and a lakeshore trail is 3.2 km in length. In June, Kathy walked the nature trail three times and the lakeshore trail twice. How many kilometers did Kathy walk on the two trails in June?

   ____________________

8. Yoko buys a pen for \$1.19 and 2 paperback books. Each book costs \$5.95. She gives the clerk a 20-dollar bill and receives \$7.91 in change. Did she receive the correct amount of change? If not, how much too little or too much did she receive?

   ____________________


Copyright © William H. Sadlier, Inc. All rights reserved.

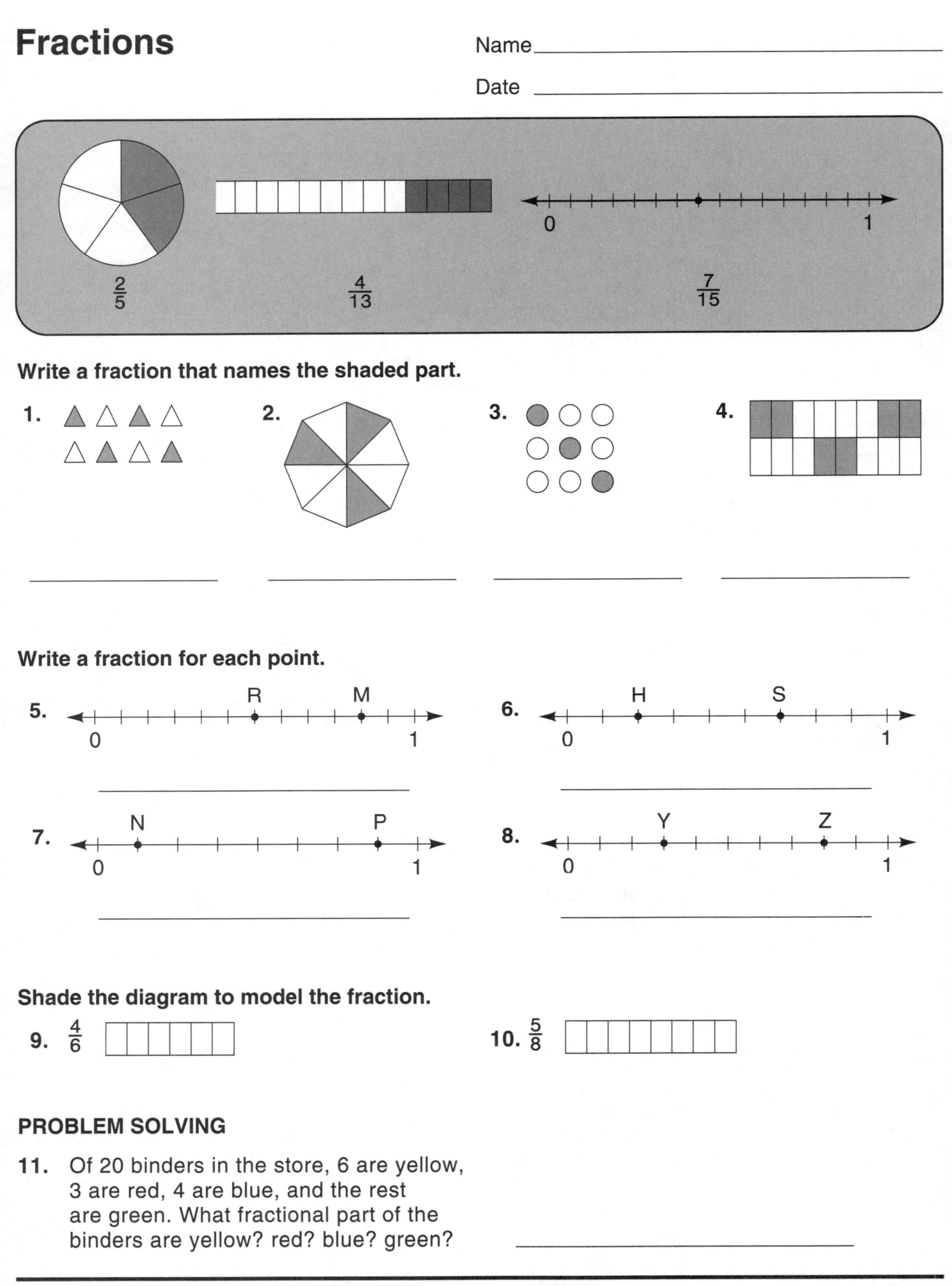

# Fractions

Name ______________________

Date ______________________

$\frac{2}{5}$ $\frac{4}{13}$ $\frac{7}{15}$

0 1

**Write a fraction that names the shaded part.**

1. ______

2. ______

3. ______

4. ______

**Write a fraction for each point.**

5. R M; 0 1 ______

6. H S; 0 1 ______

7. N P; 0 1 ______

8. Y Z; 0 1 ______

**Shade the diagram to model the fraction.**

9. $\frac{4}{6}$

10. $\frac{5}{8}$

**PROBLEM SOLVING**

11. Of 20 binders in the store, 6 are yellow, 3 are red, 4 are blue, and the rest are green. What fractional part of the binders are yellow? red? blue? green? ______

 Copyright © William H. Sadlier, Inc. All rights reserved.

# Finding Equivalent Fractions

Name ____________________

Date ____________________

$\frac{7}{8} = \frac{7 \times 4}{8 \times 4} = \frac{28}{32}$ $\frac{9}{45} = \frac{9 \div 9}{45 \div 9} = \frac{1}{5}$

$\frac{7}{8} = \frac{28}{32}$ $\frac{9}{45} = \frac{1}{5}$

**Complete.**

1. $\frac{1}{2} = \frac{\_\_}{4} = \frac{\_\_}{6}$
2. $\frac{2}{3} = \frac{\_\_}{6} = \frac{6}{\_\_}$
3. $\frac{2}{5} = \frac{\_\_}{10} = \frac{6}{\_\_}$
4. $\frac{7}{8} = \frac{14}{\_\_} = \frac{\_\_}{24}$
5. $\frac{4}{7} = \frac{8}{\_\_} = \frac{12}{\_\_}$
6. $\frac{5}{6} = \frac{\_\_}{12} = \frac{15}{\_\_}$

**Which two figures show equivalent fractions?**

7. ________ a. b. c. d.

8. ________ a. b. c. d.

9. ________ a. b. c. d.

**Write the missing number to complete the equivalent fraction.**

10. $\frac{3}{5} = \frac{12}{\_\_}$
11. $\frac{7}{12} = \frac{21}{\_\_}$
12. $\frac{1}{6} = \frac{\_\_}{18}$
13. $\frac{5}{11} = \frac{35}{\_\_}$
14. $\frac{3}{10} = \frac{33}{\_\_}$
15. $\frac{24}{56} = \frac{\_\_}{7}$
16. $\frac{3}{8} = \frac{\_\_}{40}$
17. $\frac{3}{33} = \frac{1}{\_\_}$
18. $\frac{9}{27} = \frac{1}{3}$
19. $\frac{13}{26} = \frac{1}{\_\_}$
20. $\frac{4}{9} = \frac{16}{\_\_}$
21. $\frac{5}{50} = \frac{\_\_}{10}$
22. $\frac{45}{72} = \frac{5}{\_\_}$
23. $\frac{12}{16} = \frac{3}{\_\_}$
24. $\frac{40}{55} = \frac{\_\_}{11}$
25. $\frac{56}{84} = \frac{\_\_}{12}$
26. $\frac{70}{90} = \frac{7}{\_\_}$
27. $\frac{1}{14} = \frac{\_\_}{42}$
28. $\frac{1}{5} = \frac{\_\_}{100}$
29. $\frac{35}{63} = \frac{\_\_}{9}$

Copyright © William H. Sadlier, Inc. All rights reserved.

# Prime and Composite Numbers

Name ____________________

Date ____________________

A **prime number** is a number greater than1 that has exactly two factors, itself and 1.

1 × 7 = 7 Factors of 7: 1, 7

7 is a prime number.

A **composite number** is a number greater than1 that has more than two factors.

1 × 6 = 6
2 × 3 = 6 Factors of 6: 1, 2, 3, 6

6 is a composite number.

**Write whether each is *prime*, *composite*, or *neither*.**

**1.** 6 ________ **2.** 11 ________ **3.** 15 ________

**4.** 5 ________ **5.** 38 ________ **6.** 79 ________

**7.** 1 ________ **8.** 24 ________ **9.** 56 ________

**10.** 23 ________ **11.** 2 ________ **12.** 16 ________

**13.** 90 ________ **14.** 0 ________ **15.** 3 ________

**Write *True* or *False* for each statement.**

**16.** The greatest prime number between 1and 100 is 99. ________

**17.** All even numbers are composite numbers. ________

**18.** No prime numbers are odd numbers. ________

**19.** The least prime number is 2. ________

**20.** The numbers 23, 31, 37, and 41 are all prime numbers. ________

**21.** There are no composite numbers between 70 and 80. ________

**Use the numbers in the box. Indentify the numbers.**

| 8 | 10 | 12 | 13 | 15 |
|---|---|---|---|---|
| 17 | 18 | 19 | 20 | 21 |
| 24 | 26 | 29 | 30 | |

**22.** have exactly four factors ________

**23.** have more than four factors ________

**24.** prime numbers ________

**25.** have both 2 and 4 as factors ________

**26.** have both 1 and 3 as factors ________


Copyright © William H. Sadlier, Inc. All rights reserved.

# Prime Factorization

Name ____________________

Date ____________________

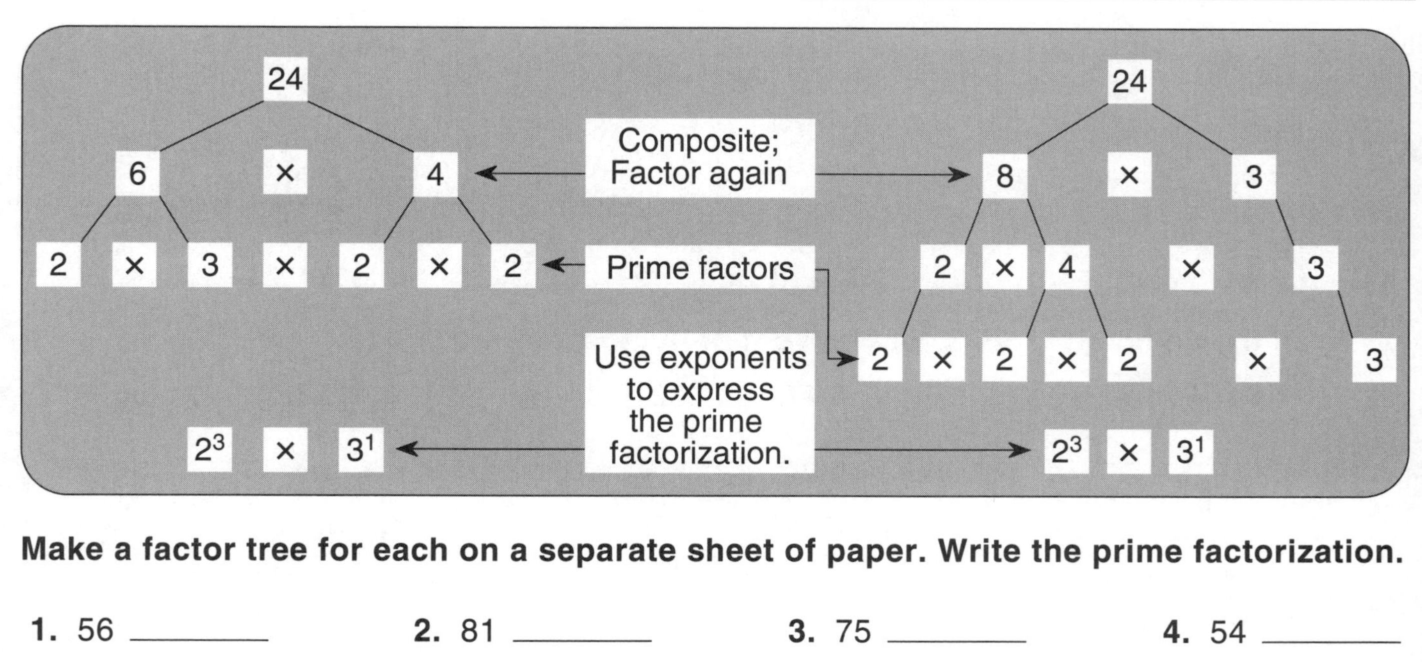


**Make a factor tree for each on a separate sheet of paper. Write the prime factorization.**

**1.** 56 ________ **2.** 81 ________ **3.** 75 ________ **4.** 54 ________

**5.** 90 ________ **6.** 84 ________ **7.** 32 ________ **8.** 63 ________

**9.** 40 ________ **10.** 55 ________ **11.** 68 ________ **12.** 92 ________

**Express each in exponent form.**

**13.** $7 \times 2 \times 7$ ________

**14.** $3 \times 3 \times 3 \times 5 \times 5$ ________

**15.** $5 \times 2 \times 5 \times 2$ ________

**16.** $3 \times 2 \times 3 \times 11 \times 2$ ________

**17.** $2 \times 2 \times 3 \times 3 \times 2$ ________

**18.** $13 \times 3 \times 3 \times 13 \times 2$ ________

**Find the prime factorization. Use divisibility rules.**

**19.** 98 ________ **20.** 279 ________ **21.** 560 ________ **22.** 225 ________

**23.** 490 ________ **24.** 728 ________ **25.** 405 ________ **26.** 1665 ________

Copyright © William H. Sadlier, Inc. All rights reserved.

# Greatest Common Factor

Name ______________________

Date ______________________

Find the greatest common factor of 6, 9, and 12.
List the factors of each number.

6 : **1**, 2, **3**, 6
9 : **1**, **3**, 9
12 : **1**, 2, **3**, 4, 6, 12

**Common Factors:** 1 and 3
**Greatest Common Factor:** 3

**Write all the common factors for each set of numbers.**

1. 10 and 45 ________
2. 12 and 32 ________
3. 24 and 36 ________
4. 9 and 54 ________
5. 15 and 75 ________
6. 60 and 24 ________
7. 4, 8, and 16 ________
8. 7, 13, and 23 ________
9. 8, 20, and 34 ________

**Find the GCF for each set of numbers.**

10. 8 and 16 ________
11. 6 and 10 ________
12. 12 and 20 ________
13. 14 and 42 ________
14. 7 and 35 ________
15. 9 and 15 ________
16. 5 and 30 ________
17. 8 and 25 ________
18. 7, 28, and 35 ________
19. 4, 12, and 24 ________
20. 13, 26, and 65 ________
21. 18, 51, and 81 ________

**Find the GCF. Use prime factorization.**

22. 36 and 60 ________
23. 40 and 65 ________
24. 28 and 63 ________
25. 70 and 105 ________
26. 26 and 91 ________
27. 33 and 75 ________
28. 56 and 84 ________
29. 81 and 99 ________
30. 15, 45, and 60 ________
31. 34, 85, and 102 ________
32. 16, 64, and 96 ________
33. 56, 64, and 104 ________

**Find the pairs of numbers.**

34. between 8 and 18 that have 5 as their GCF ________
35. between 15 and 30 that have 7 as their GCF ________


Copyright © William H. Sadlier, Inc. All rights reserved.

# Fractions in Simplest Form

Name ____________________

Date ____________________

Rename $\frac{21}{56}$ in simplest form, or lowest terms.

Find the GCF of the numerator and the denominator.

Divide the numerator and the denominator by their GCF.

21 : 1, 3, 7, 21

56 : 1, 2, 4, 7, 8, 14, 28, 56

$\frac{21}{56} = \frac{21 \div 7}{56 \div 7} = \frac{3}{8}$

The simplest form of $\frac{21}{56}$ is $\frac{3}{8}$.

**Circle the letter of the GCF of the numerator and the denominator of each fraction.**

| | | | | |
|---|---|---|---|---|
| **1.** $\frac{9}{12}$ | **a.** 6 | **b.** 3 | **c.** 2 | **d.** 9 |
| **2.** $\frac{4}{8}$ | **a.** 1 | **b.** 2 | **c.** 0 | **d.** 4 |
| **3.** $\frac{6}{15}$ | **a.** 3 | **b.** 6 | **c.** 5 | **d.** 12 |
| **4.** $\frac{16}{40}$ | **a.** 2 | **b.** 4 | **c.** 8 | **d.** 16 |
| **5.** $\frac{20}{70}$ | **a.** 7 | **b.** 2 | **c.** 10 | **d.** 20 |
| **6.** $\frac{45}{63}$ | **a.** 5 | **b.** 9 | **c.** 15 | **d.** 30 |

**Is the fraction in lowest terms? Write *Yes* or *No*.**

**7.** $\frac{3}{8}$ ___ **8.** $\frac{4}{10}$ ___ **9.** $\frac{7}{12}$ ___ **10.** $\frac{5}{16}$ ___ **11.** $\frac{2}{4}$ ___ **12.** $\frac{12}{27}$ ___

**13.** $\frac{14}{20}$ ___ **14.** $\frac{8}{32}$ ___ **15.** $\frac{11}{33}$ ___ **16.** $\frac{3}{18}$ ___ **17.** $\frac{5}{12}$ ___ **18.** $\frac{13}{20}$ ___

**Rename each as a fraction in simplest form. Do your work on a separate sheet of paper.**

**19.** $\frac{2}{8}$ ___ **20.** $\frac{10}{15}$ ___ **21.** $\frac{21}{28}$ ___ **22.** $\frac{35}{40}$ ___ **23.** $\frac{6}{15}$ ___ **24.** $\frac{18}{81}$ ___

**25.** $\frac{11}{55}$ ___ **26.** $\frac{15}{18}$ ___ **27.** $\frac{28}{56}$ ___ **28.** $\frac{9}{24}$ ___ **29.** $\frac{9}{30}$ ___ **30.** $\frac{24}{72}$ ___

**31.** $\frac{28}{32}$ ___ **32.** $\frac{6}{42}$ ___ **33.** $\frac{8}{88}$ ___ **34.** $\frac{15}{25}$ ___ **35.** $\frac{14}{63}$ ___ **36.** $\frac{5}{60}$ ___

**37.** $\frac{4}{24}$ ___ **38.** $\frac{10}{16}$ ___ **39.** $\frac{21}{36}$ ___ **40.** $\frac{36}{42}$ ___ **41.** $\frac{32}{40}$ ___ **42.** $\frac{40}{72}$ ___

**PROBLEM SOLVING Write the answer as a fraction in simplest form.**

**43.** The movie theater sold 28 matinee tickets and 56 tickets to the evening show. What part of the tickets sold were for the matinee? ____________

Copyright © William H. Sadlier, Inc. All rights reserved.

# Mixed Numbers and Improper Fractions

Name ______________________

Date ______________________

Rename $2\frac{1}{4}$ as an improper fraction.

$$2\frac{1}{4} = \frac{(4 \times 2) + 1}{4} = \frac{9}{4}$$

Rename $\frac{30}{8}$ as a mixed number.

$$\frac{30}{8} = 8\overline{)30} \quad 3 \text{ R}6$$

$$\frac{30}{8} = 3\frac{6}{8}$$

$$= 3\frac{3}{4}$$

**Write the word name for each mixed number.**

1. $4\frac{2}{3}$ ______________________
2. $1\frac{1}{2}$ ______________________
3. $38\frac{5}{12}$ ______________________
4. $16\frac{7}{10}$ ______________________

**Write as a mixed number.**

5. two and four fifths ____________
6. six and nine tenths ____________
7. three and nine twentieths ____________
8. eleven and eight fifteenths ____________

**Express each as an improper fraction.**

9. $1\frac{5}{8}$ ________
10. $6\frac{5}{6}$ ________
11. $13\frac{1}{2}$ ________
12. $5\frac{1}{3}$ ________
13. $7\frac{3}{4}$ ________
14. $8\frac{5}{7}$ ________
15. $10\frac{1}{8}$ ________
16. $2\frac{2}{5}$ ________
17. $3\frac{1}{6}$ ________
18. $9\frac{3}{10}$ ________
19. $4\frac{3}{4}$ ________
20. $11\frac{7}{8}$ ________
21. $6\frac{2}{9}$ ________
22. $10\frac{3}{4}$ ________
23. $8\frac{3}{5}$ ________
24. $3\frac{5}{12}$ ________

**Express each as a whole number or a mixed number in simplest form.**

25. $\frac{27}{4}$ ________
26. $\frac{94}{8}$ ________
27. $\frac{51}{6}$ ________
28. $\frac{25}{10}$ ________
29. $\frac{26}{5}$ ________
30. $\frac{84}{7}$ ________
31. $\frac{9}{2}$ ________
32. $\frac{48}{3}$ ________
33. $\frac{7}{2}$ ________
34. $\frac{48}{5}$ ________
35. $\frac{32}{9}$ ________
36. $\frac{64}{40}$ ________
37. $\frac{22}{4}$ ________
38. $\frac{66}{8}$ ________
39. $\frac{39}{3}$ ________
40. $\frac{72}{10}$ ________

 Use with Lesson 5-7, text pages 174–175. Copyright © William H. Sadlier, Inc. All rights reserved.

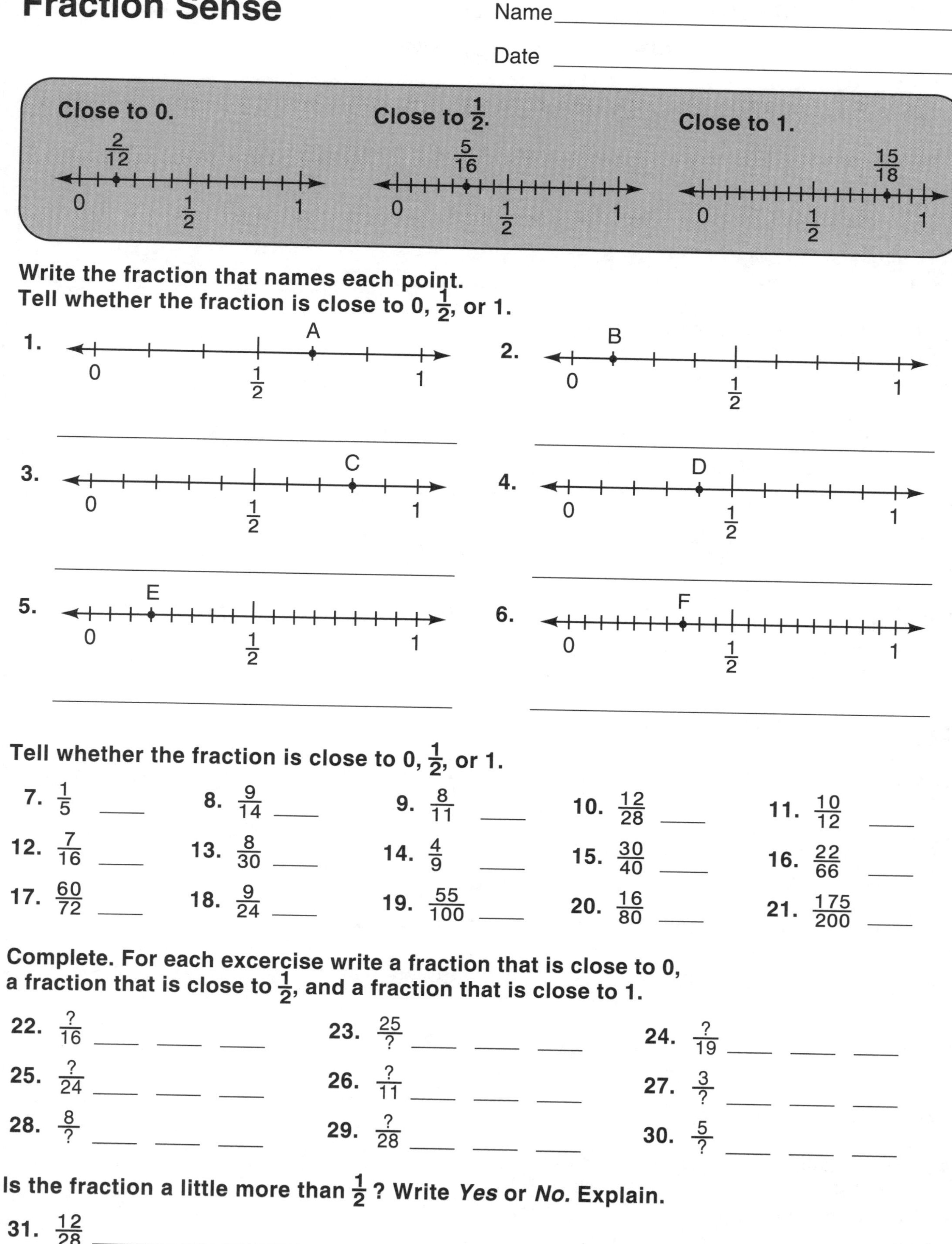

# Fraction Sense

Name ____________________

Date ____________________

**Close to 0.**

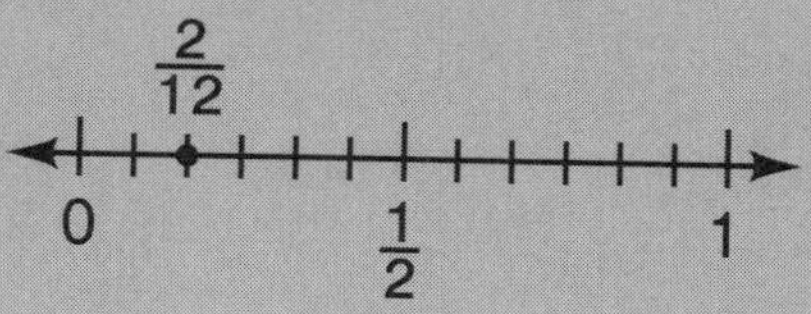


**Close to $\frac{1}{2}$.**

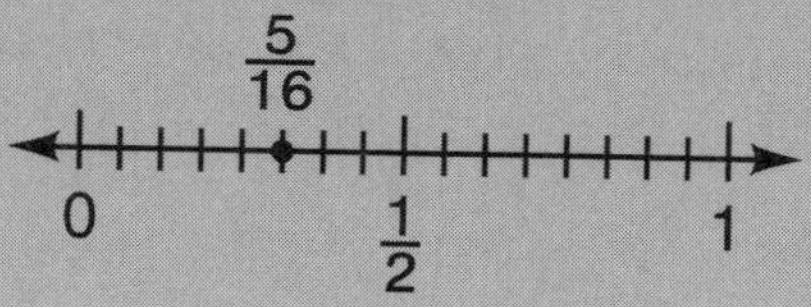


**Close to 1.**

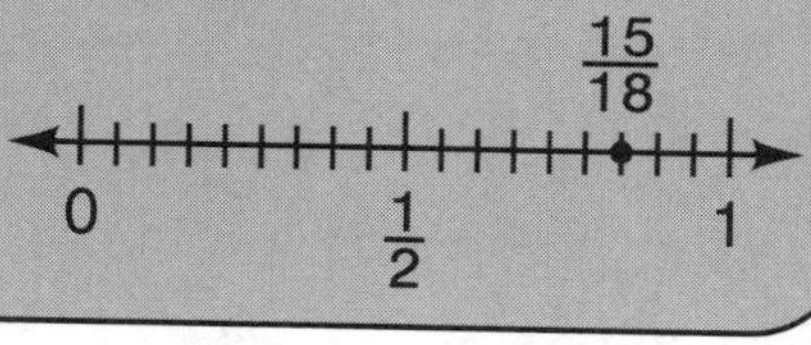


**Write the fraction that names each point.**
**Tell whether the fraction is close to 0, $\frac{1}{2}$, or 1.**

**1.**

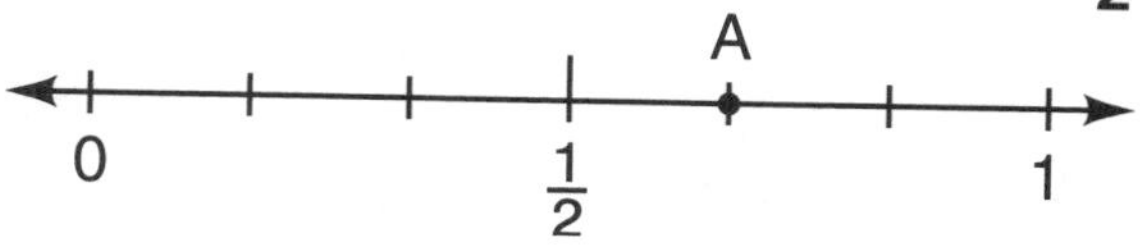


____________________

**2.**

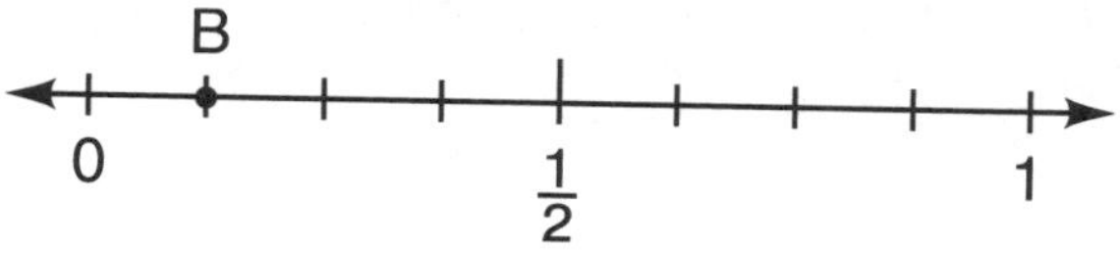


____________________

**3.**

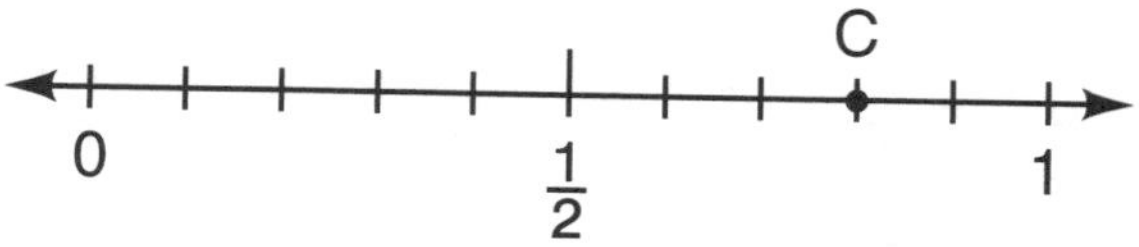


____________________

**4.**

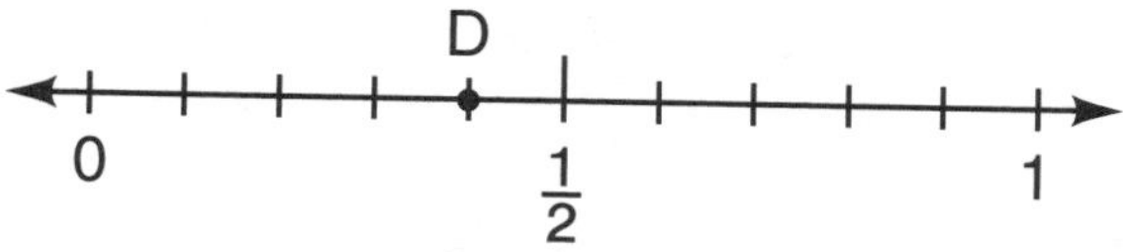


____________________

**5.**

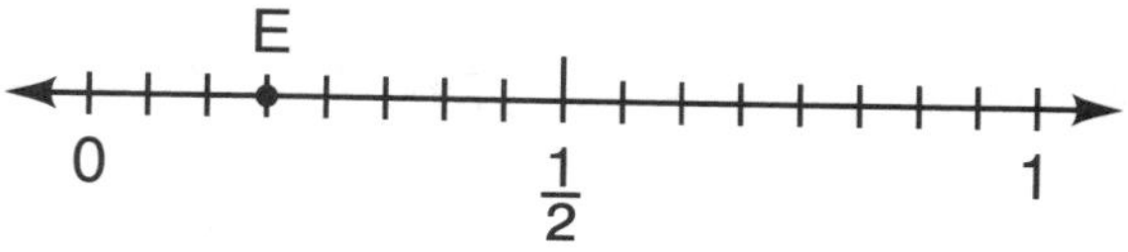


____________________

**6.**

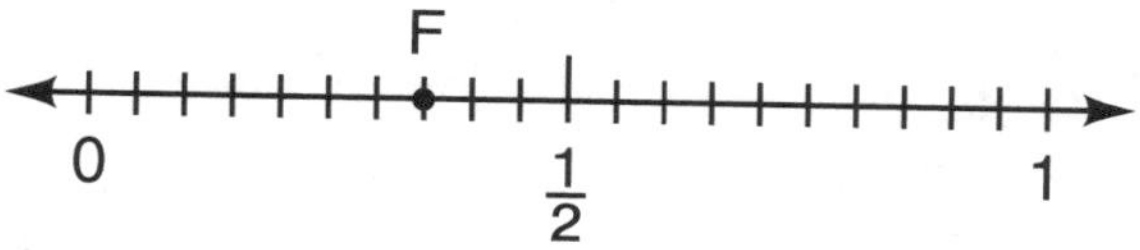


____________________

**Tell whether the fraction is close to 0, $\frac{1}{2}$, or 1.**

**7.** $\frac{1}{5}$ ____ **8.** $\frac{9}{14}$ ____ **9.** $\frac{8}{11}$ ____ **10.** $\frac{12}{28}$ ____ **11.** $\frac{10}{12}$ ____

**12.** $\frac{7}{16}$ ____ **13.** $\frac{8}{30}$ ____ **14.** $\frac{4}{9}$ ____ **15.** $\frac{30}{40}$ ____ **16.** $\frac{22}{66}$ ____

**17.** $\frac{60}{72}$ ____ **18.** $\frac{9}{24}$ ____ **19.** $\frac{55}{100}$ ____ **20.** $\frac{16}{80}$ ____ **21.** $\frac{175}{200}$ ____

**Complete. For each excercise write a fraction that is close to 0, a fraction that is close to $\frac{1}{2}$, and a fraction that is close to 1.**

**22.** $\frac{?}{16}$ ____ ____ ____ **23.** $\frac{25}{?}$ ____ ____ ____ **24.** $\frac{?}{19}$ ____ ____ ____

**25.** $\frac{?}{24}$ ____ ____ ____ **26.** $\frac{?}{11}$ ____ ____ ____ **27.** $\frac{3}{?}$ ____ ____ ____

**28.** $\frac{8}{?}$ ____ ____ ____ **29.** $\frac{?}{28}$ ____ ____ ____ **30.** $\frac{5}{?}$ ____ ____ ____

**Is the fraction a little more than $\frac{1}{2}$? Write *Yes* or *No*. Explain.**

**31.** $\frac{12}{28}$ ________________________________________

Copyright © William H. Sadlier, Inc. All rights reserved.

# Least Common Multiple

Name ______________________

Date ______________________

What is the LCM of 4, 6, and 8?

| **Multiples** | **Least Common Multiple (LCM)** |
|---|---|
| 4 : 4, 8, 12, 16, 20, **24** . . . | 24 |
| 6 : 6, 12, 18, **24**, 30 . . . | |
| 8 : 8, 16, **24**, 32 . . . | |

**Find the LCM of each set of numbers.**

**1.** 4, 6 _____ **2.** 3, 5 _____ **3.** 6, 12 _____ **4.** 5, 7 _____

**5.** 4, 24 _____ **6.** 6, 15 _____ **7.** 7, 11 _____ **8.** 9, 30 _____

**9.** 5, 8 _____ **10.** 1, 17 _____ **11.** 8, 48 _____ **12.** 3, 12 _____

**13.** 32, 48 _____ **14.** 2, 35 _____ **15.** 10, 12 _____ **16.** 16, 64 _____

**17.** 7, 8 _____ **18.** 14, 42 _____ **19.** 9, 36 _____ **20.** 11, 12 _____

**21.** 5, 10, and 15 _____ **22.** 3, 6, and 9 _____ **23.** 4, 8, and 12 _____

**24.** 2, 3, and 5 _____ **25.** 3, 4, and 6 _____ **26.** 4, 6, and 9 _____

**27.** 3, 5, and 9 _____ **28.** 3, 4, and 7 _____ **29.** 4, 7, and 8 _____

**30.** 4, 6, and 32 _____ **31.** 5, 8, and 20 _____ **32.** 3, 4, and 18 _____

**Use a shortcut to find the LCM of each pair of numbers.**

**33.** 3, 7 _____ **34.** 2, 16 _____ **35.** 3, 9 _____ **36.** 5, 9 _____

**37.** 5, 25 _____ **38.** 4, 12 _____ **39.** 7, 11 _____ **40.** 2, 13 _____

**PROBLEM SOLVING**

**41.** Tasha lists all the multiples of 3 from 3 to 99. Tony lists all the multiples of 5 from 5 to 100. What is the first number that is on both lists? the second? __________


Copyright © William H. Sadlier, Inc. All rights reserved.

# Comparing Fractions

Name ____________________

Date ____________________

| Compare: $\frac{7}{12} \underline{\ ?\ } \frac{11}{12}$ | Compare: $\frac{2}{3} \underline{\ ?\ } \frac{3}{5}$ |
|---|---|
| $\frac{7}{12} \underline{\ ?\ } \frac{11}{12} \longrightarrow 7 < 11$ | $\frac{2}{3} \underline{\ ?\ } \frac{3}{5}$ LCD is 15. |
| So $\frac{7}{12} < \frac{11}{12}$. | $\frac{2}{3} = \frac{?}{15} \longrightarrow \frac{2 \times 5}{3 \times 5} = \frac{10}{15}$ |
| | $\frac{3}{5} = \frac{?}{15} \longrightarrow \frac{3 \times 3}{5 \times 3} = \frac{9}{15}$ $10 > 9$ |
| | So $\frac{10}{15} > \frac{9}{15}$ and $\frac{2}{3} > \frac{3}{5}$. |

**Compare. Write <, =, or >.**

1. $\frac{2}{5}$ ___ $\frac{3}{5}$
2. $\frac{3}{8}$ ___ $\frac{7}{8}$
3. $\frac{11}{12}$ ___ $\frac{11}{12}$
4. $\frac{9}{10}$ ___ $\frac{4}{10}$
5. $\frac{14}{25}$ ___ $\frac{21}{25}$
6. $\frac{5}{7}$ ___ $\frac{1}{7}$
7. $\frac{21}{28}$ ___ $\frac{24}{28}$
8. $\frac{17}{18}$ ___ $\frac{7}{18}$
9. $\frac{20}{30}$ ___ $\frac{25}{30}$
10. $\frac{9}{11}$ ___ $\frac{5}{11}$

**Write the LCD of each pair of fractions. Then rename the fractions so they have the LCD as their denominator.**

11. $\frac{3}{4}, \frac{2}{12}$ ____________
12. $\frac{7}{8}, \frac{1}{2}$ ____________
13. $\frac{2}{5}, \frac{3}{10}$ ____________
14. $\frac{1}{4}, \frac{1}{5}$ ____________
15. $\frac{1}{2}, \frac{3}{7}$ ____________
16. $\frac{3}{8}, \frac{5}{12}$ ____________
17. $\frac{5}{9}, \frac{1}{8}$ ____________
18. $\frac{2}{3}, \frac{9}{10}$ ____________
19. $\frac{4}{5}, \frac{1}{18}$ ____________

**Compare. Write <, =, or >.**

20. $\frac{3}{4}$ ___ $\frac{3}{7}$
21. $\frac{7}{8}$ ___ $\frac{2}{5}$
22. $\frac{1}{4}$ ___ $\frac{3}{10}$
23. $\frac{2}{3}$ ___ $\frac{8}{12}$
24. $\frac{1}{2}$ ___ $\frac{5}{12}$
25. $\frac{4}{5}$ ___ $\frac{7}{9}$
26. $\frac{9}{12}$ ___ $\frac{3}{8}$
27. $\frac{14}{30}$ ___ $\frac{7}{15}$
28. $\frac{2}{5}$ ___ $\frac{15}{25}$

**PROBLEM SOLVING**

29. Mrs. Johnson bought $\frac{5}{8}$ yd of gingham and $\frac{2}{3}$ yd of calico. Did she buy more gingham or more calico? ____________

Copyright © William H. Sadlier, Inc. All rights reserved.

# Ordering Fractions

Name ______________________________

Date ______________________________

**Order from least to greatest:**

$\frac{3}{8}, \frac{7}{12}, \frac{1}{3}$ LCD is 24.

$\frac{3}{8} = \frac{9}{24}, \frac{7}{12} = \frac{14}{24}, \frac{1}{3} = \frac{8}{24}$

$\frac{8}{24} < \frac{9}{24} < \frac{14}{24}$

**From greatest to least:**

$\frac{1}{3}, \frac{3}{8}, \frac{7}{12}$

**Order from greatest to least:**

$3\frac{1}{3}, 3\frac{9}{15}, \frac{17}{5}$ $\frac{17}{5} = 3\frac{2}{5}$ LCD is 15.

$3\frac{5}{15}, 3\frac{9}{15}, 3\frac{6}{15}$

$3\frac{9}{15} > 3\frac{6}{15} > 3\frac{5}{15}$

**From greatest to least:**

$3\frac{9}{15}, \frac{17}{5}, 3\frac{1}{3}$

**Write in order from least to greatest.**

1. $\frac{8}{12}, \frac{6}{12}, \frac{7}{12}$ ______________
2. $\frac{9}{15}, \frac{11}{15}, \frac{6}{15}$ ______________
3. $\frac{3}{4}, \frac{2}{8}, \frac{5}{8}$ ______________
4. $\frac{5}{9}, \frac{1}{3}, \frac{2}{9}$ ______________
5. $\frac{2}{7}, \frac{1}{2}, \frac{3}{5}$ ______________
6. $\frac{1}{10}, \frac{1}{9}, \frac{1}{5}$ ______________
7. $\frac{3}{4}, \frac{5}{7}, \frac{7}{8}$ ______________
8. $\frac{3}{10}, \frac{1}{4}, \frac{2}{5}$ ______________
9. $5\frac{2}{3}, 5\frac{5}{9}, 5\frac{11}{15}$ ______________
10. $1\frac{1}{3}, 1\frac{3}{5}, 1\frac{3}{10}$ ______________
11. $7\frac{5}{12}, 7\frac{1}{2}, 7\frac{3}{8}$ ______________
12. $4\frac{9}{10}, 4\frac{3}{4}, 4\frac{5}{8}$ ______________

**Write in order from greatest to least.**

13. $\frac{3}{8}, \frac{7}{8}, \frac{5}{8}$ ______________
14. $\frac{7}{10}, \frac{3}{10}, \frac{9}{10}$ ______________
15. $\frac{6}{15}, \frac{4}{5}, \frac{3}{5}$ ______________
16. $\frac{9}{12}, \frac{1}{4}, \frac{5}{12}$ ______________
17. $\frac{11}{12}, \frac{7}{8}, \frac{5}{6}$ ______________
18. $\frac{3}{7}, \frac{1}{4}, \frac{5}{14}$ ______________
19. $\frac{4}{9}, \frac{1}{3}, \frac{1}{2}$ ______________
20. $\frac{2}{3}, \frac{8}{15}, \frac{7}{12}$ ______________
21. $8\frac{3}{10}, 8\frac{7}{20}, 8\frac{2}{5}$ ______________
22. $\frac{31}{7}, \frac{13}{14}, \frac{21}{7}$ ______________
23. $1\frac{3}{9}, 1\frac{2}{5}, 1\frac{1}{2}$ ______________
24. $4\frac{1}{4}, \frac{38}{8}, 4\frac{1}{2}$ ______________

**PROBLEM SOLVING**

25. Recipe A calls for $\frac{2}{3}$ c cornmeal, recipe B calls for $\frac{5}{8}$ c cornmeal, and recipe C calls for $\frac{1}{2}$ c cornmeal. Which recipe uses the most cornmeal? ______________

 Copyright © William H. Sadlier, Inc. All rights reserved.

# Fractions, Mixed Numbers, Decimals

Name ____________________

Date ____________________

| thirteen hundredths | | one and two tenths | | sixty-one thousandths | |
|---|---|---|---|---|---|
| $\frac{13}{100}$ | 0.13 | $1\frac{2}{10}$ | 1.2 | $\frac{61}{1000}$ | 0.061 |

**Write the letter of the equivalent decimal.**

1. $\frac{9}{10}$ ____ **a.** 0.9 **b.** 0.09 **c.** 0.009 **d.** 9
2. $8\frac{17}{100}$ ____ **a.** 0.817 **b.** 0.0817 **c.** 8.17 **d.** 817
3. $\frac{466}{1000}$ ____ **a.** 466.000 **b.** 4.0066 **c.** 4.66 **d.** 0.466

**Write the letter of the equivalent fraction.**

4. 0.012 ____ **a.** $\frac{12}{10}$ **b.** $\frac{12}{100}$ **c.** $\frac{12}{1000}$ **d.** $\frac{12}{10,000}$
5. 2.6139 ____ **a.** $\frac{26139}{10}$ **b.** $2\frac{6139}{1000}$ **c.** $26\frac{139}{100}$ **d.** $2\frac{6139}{10,000}$
6. 0.75 ____ **a.** $\frac{75}{100}$ **b.** $\frac{750}{1000}$ **c.** $\frac{75}{10}$ **d.** $7\frac{5}{10}$

**Write the word name. Then write the equivalent decimal or fraction.**

7. $8\frac{8}{100}$ ______________________________ ________
8. $16\frac{114}{1000}$ ______________________________ ________
9. 65.22 ______________________________ ________
10. 12.0037 ______________________________ ________
11. 0.005 ______________________________ ________
12. 0.0400 ______________________________ ________

**Write the equivalent decimal or whole number.**

13. $\frac{44}{10}$ ____ 14. $\frac{190}{100}$ ____ 15. $\frac{560}{10}$ ____ 16. $\frac{70}{10}$ ____

17. $\frac{2550}{1000}$ ____ 18. $\frac{399}{100}$ ____ 19. $\frac{859}{100}$ ____ 20. $\frac{1500}{1000}$ ____

21. $\frac{6124}{1000}$ ____ 22. $\frac{18,123}{10,000}$ ____ 23. $\frac{95,000}{10,000}$ ____ 24. $\frac{42,500}{10,000}$ ____

Copyright © William H. Sadlier, Inc. All rights reserved.

**Use with Lesson 5-12, text pages 184–185.**

# Fractions: Renaming as Decimals

Name ____________________

Date ____________________

| Rename $\frac{5}{8}$ as a decimal. | Rename $3\frac{1}{4}$ as a decimal. |
|---|---|
| $8\overline{)5.000}$ = 0.625 | $3\frac{1}{4} = 3 + \frac{1}{4}$ |
| | $\frac{1}{4} \longrightarrow 4\overline{)1.00}$ = 0.25 |
| So $\frac{5}{8} = 0.625$ | $3 + 0.25 = 3.25$ |
| | So $3\frac{1}{4} = 3.25$ |

**Write each fraction as a decimal.**

**1.** $\frac{1}{2}$ ______ **2.** $\frac{3}{20}$ ______ **3.** $\frac{4}{50}$ ______ **4.** $\frac{6}{8}$ ______

**5.** $\frac{6}{25}$ ______ **6.** $\frac{12}{15}$ ______ **7.** $\frac{3}{4}$ ______ **8.** $\frac{35}{40}$ ______

**9.** $\frac{4}{5}$ ______ **10.** $\frac{3}{8}$ ______ **11.** $\frac{3}{6}$ ______ **12.** $\frac{1}{16}$ ______

**13.** $\frac{6}{24}$ ______ **14.** $\frac{45}{75}$ ______ **15.** $\frac{54}{72}$ ______ **16.** $\frac{14}{20}$ ______

**Write each mixed number as a decimal.**

**17.** $2\frac{1}{4}$ ______ **18.** $19\frac{3}{5}$ ______ **19.** $8\frac{2}{25}$ ______ **20.** $22\frac{3}{20}$ ______

**21.** $15\frac{1}{8}$ ______ **22.** $30\frac{12}{50}$ ______ **23.** $61\frac{3}{10}$ ______ **24.** $76\frac{4}{5}$ ______

**25.** $45\frac{15}{20}$ ______ **26.** $17\frac{4}{10}$ ______ **27.** $51\frac{9}{15}$ ______ **28.** $10\frac{86}{100}$ ______

**29.** $1\frac{2}{5}$ ______ **30.** $5\frac{7}{8}$ ______ **31.** $16\frac{3}{40}$ ______ **32.** $9\frac{17}{50}$ ______

**33.** $20\frac{3}{16}$ ______ **34.** $3\frac{7}{25}$ ______ **35.** $7\frac{7}{50}$ ______ **36.** $34\frac{9}{20}$ ______

**PROBLEM SOLVING**

**37.** Julio has six twelfths of a dollar. How much money does he have? ____________________

**38.** Lavonne has six and seven twentieths dollars. How much money does she have? ____________________

 **Use with Lesson 5-13, text pages 185–186.** Copyright © William H. Sadlier, Inc. All rights reserved.

# Decimals as Fractions and Mixed Numbers

Name ____________________

Date ____________________

Write 0.44 as a fraction in simplest form.

$0.44 = \frac{44}{100}$

$\frac{44}{100} = \frac{44 \div 4}{100 \div 4} = \frac{11}{25}$

So $0.44 = \frac{11}{25}$.

Write 6.002 as a mixed number in simplest form.

$6.002 = 6\frac{2}{1000}$

$\frac{2}{1000} = \frac{2 \div 2}{1000 \div 2} = \frac{1}{500}$

So $6.002 = 6\frac{1}{500}$.

**Complete.**

1. $0.8 = \frac{8}{\phantom{0}} = \frac{\phantom{0}}{5}$
2. $0.24 = \frac{24}{\phantom{0}} = \frac{\phantom{0}}{25}$
3. $0.009 = \frac{\phantom{0}}{1000}$
4. $3.25 = 3\frac{25}{\phantom{0}} = 3\frac{\phantom{0}}{4}$
5. $8.063 = 8\frac{\phantom{0}}{1000}$
6. $6.875 = 6$ ____

**Write each decimal as a fraction in simplest form.**

7. 0.3 ________
8. 0.17 ________
9. 0.009 ________
10. 0.387 ________
11. 0.125 ________
12. 0.0123 ________
13. 0.62 ________
14. 0.046 ________
15. 0.4 ________
16. 0.12 ________
17. 0.275 ________
18. 0.0025 ________
19. 0.099 ________
20. 0.5 ________
21. 0.48 ________
22. 0.0125 ________

**Write each decimal as a mixed number in simplest form.**

23. 5.04 ________
24. 9.12 ________
25. 8.133 ________
26. 6.01 ________
27. 7.625 ________
28. 2.25 ________
29. 1.325 ________
30. 10.6 ________
31. 3.08 ________
32. 2.0004 ________
33. 5.0125 ________
34. 9.42 ________
35. The newborn baby weighed 3.2 kilograms. ________
36. The newborn calf weighed 14.25 pounds. ________


Copyright © William H. Sadlier, Inc. All rights reserved.

# Terminating and Repeating Decimals

Name ______________________

Date ______________________

| Terminating Decimals | Repeating Decimals |
|---|---|
| $\frac{1}{4} \rightarrow 4\overline{)1.00}$ = 0.25 | $\frac{1}{6} \rightarrow 6\overline{)1.00000}$ = 0.16666 . . . $\rightarrow 0.1\overline{6}$ |
| $\frac{5}{16} \rightarrow 16\overline{)5.0000}$ = 0.3125 | $\frac{3}{11} \rightarrow 11\overline{)3.000000}$ = 0.272727 . . . $\rightarrow 0.\overline{27}$ |

**Rewrite each repeating decimal with a bar over the part that repeats.**

1. 0.12121 . . . ________
2. 0.9166 . . . ________
3. 0.1818 . . . ________
4. 2.54545 . . . ________
5. 5.3888 . . . ________
6. 1.2666 . . . ________

**Write each repeating decimal showing eight decimal places.**

7. $0.\overline{6}$ ________
8. $0.\overline{27}$ ________
9. $0.\overline{3125}$ ________
10. $4.0\overline{9}$ ________
11. $7.\overline{83}$ ________
12. $11.\overline{545}$ ________
13. $6.1\overline{6}$ ________
14. $9.\overline{1}$ ________
15. $2.08\overline{3}$ ________

**Rename each fraction as a terminating or repeating decimal.**

16. $\frac{1}{4}$ ________
17. $\frac{2}{3}$ ________
18. $\frac{4}{5}$ ________
19. $\frac{9}{11}$ ________
20. $\frac{7}{20}$ ________
21. $\frac{4}{10}$ ________
22. $\frac{5}{9}$ ________
23. $\frac{9}{25}$ ________
24. $\frac{7}{18}$ ________
25. $\frac{5}{16}$ ________
26. $\frac{7}{12}$ ________
27. $\frac{13}{20}$ ________
28. $\frac{1}{18}$ ________
29. $\frac{4}{9}$ ________
30. $\frac{5}{8}$ ________

**Rename each mixed number as a terminating or repeating decimal.**

31. $2\frac{1}{3}$ ________
32. $5\frac{1}{8}$ ________
33. $9\frac{1}{2}$ ________
34. $14\frac{5}{12}$ ________
35. $3\frac{3}{4}$ ________
36. $28\frac{1}{25}$ ________
37. $41\frac{2}{3}$ ________
38. $35\frac{2}{9}$ ________
39. $19\frac{3}{16}$ ________
40. $13\frac{5}{8}$ ________
41. $100\frac{16}{25}$ ________
42. $64\frac{7}{12}$ ________
43. $8\frac{5}{18}$ ________
44. $7\frac{27}{36}$ ________
45. $99\frac{44}{100}$ ________
46. $58\frac{16}{24}$ ________

 Copyright © William H. Sadlier, Inc. All rights reserved.

# Problem-Solving Strategy: Find a Pattern

Name ____________________

Date ____________________

There are 8 people in a room. If each person shakes hands with each other person in the room, how many handshakes will there be?

Make a table and look for a pattern.

| People | 2 | 3 | 4 | 5 | 6 | 7 | 8 |
|---|---|---|---|---|---|---|---|
| Handshakes | 1 | 3 | 6 | 10 | 15 | 21 | 28 |

+2 +3 +4 +5 +6 +7

There will be 28 handshakes.

**PROBLEM SOLVING Do your work on a separate piece of paper.**

1. Luella has designed a pattern like the one shown. She wants to add more squares of dots around the pattern. If she continues the pattern for two more squares, how many dots will be in the largest square?

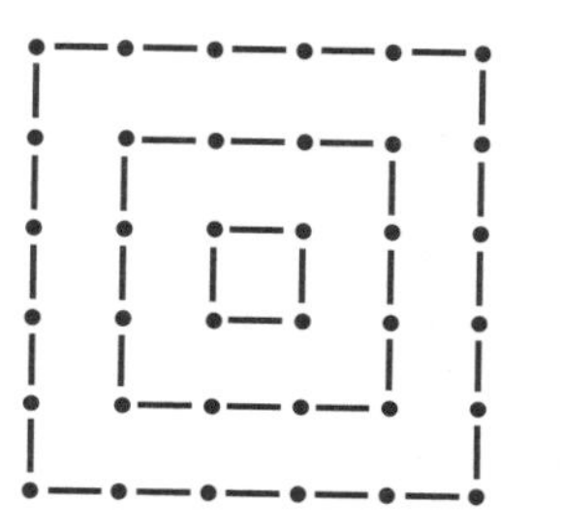

____________________

2. Ellis uses tiles to make the figures at the right. If he continues the pattern, how many tiles will he need to make the sixth figure?

First Second Third

____________________

3. Write the next three terms in this sequence: 0.1, 0.3, 0.5, 0.7. What is the pattern?

____________________

4. There are 12 basketball teams in the league. Each team will play 1 game with each of the other teams. How many games will there be?

____________________

5. Don saves $1 the first week, $2 the second week, $3 the third week, and $4 the fourth week. If he continues in this way, how many weeks will it take before Don has saved at least $50?

____________________

6. Darryl is placing colored tiles around a swimming pool. He has started the pattern with red, white, blue, white, and red and then repeated the same color sequence. If he continues this pattern, what will be the color of the 43rd tile?

____________________


Copyright © William H. Sadlier, Inc. All rights reserved.

# Addition Properties: Fractions

Name ______________________________

Date ______________________________

**Commutative**

$\frac{1}{6} + \frac{4}{6} = \frac{4}{6} + \frac{1}{6}$

$\frac{5}{6} = \frac{5}{6}$

**Associative**

$\left(\frac{1}{8} + \frac{2}{8}\right) + \frac{4}{8} = \frac{1}{8} + \left(\frac{2}{8} + \frac{4}{8}\right)$

$\frac{3}{8} + \frac{4}{8} = \frac{1}{8} + \frac{6}{8}$

$\frac{7}{8} = \frac{7}{8}$

**Identity**

$\frac{1}{2} + 0 = \frac{1}{2}$ or $0 + \frac{1}{2} = \frac{1}{2}$

**Complete.**

**1.** $\frac{1}{5} + \frac{3}{5} = \frac{3}{5} + \frac{}{5}$

$\frac{}{5} = \frac{}{5}$

**2.** $\left(\frac{1}{9} + \frac{3}{9}\right) + \frac{4}{9} = \frac{1}{9} + \left(\frac{}{9} + \frac{}{9}\right)$

$\frac{}{9} + \frac{4}{9} = \frac{1}{9} + \frac{}{9}$

$\frac{}{9} = \frac{}{9}$

**3.** $0 + \frac{3}{4} =$ ______

**4.** $\left(\frac{2}{8} + \frac{1}{8}\right) + \frac{}{8} = \frac{2}{8} + \left(\frac{1}{8} + \frac{4}{8}\right)$

$\frac{}{8} + \frac{}{8} = \frac{2}{8} + \frac{}{8}$

$\frac{}{8} = \frac{}{8}$

**5.** $\frac{2}{3} + 0 =$ ______

**6.** $\frac{3}{10} + \frac{4}{10} = \frac{4}{10} + \frac{}{10}$

$\frac{}{10} = \frac{}{10}$

**Circle the letter of the sum in simplest form.**

| | | a. | b. | c. | d. |
|---|---|---|---|---|---|
| **7.** | $\frac{3}{8} + \frac{1}{8}$ or $\frac{1}{8} + \frac{3}{8}$ | $\frac{4}{8}$ | $\frac{1}{2}$ | $\frac{1}{4}$ | $\frac{4}{16}$ |
| **8.** | $\frac{2}{5} + \frac{4}{5}$ or $\frac{4}{5} + \frac{2}{5}$ | $\frac{7}{5}$ | $\frac{7}{10}$ | $1\frac{2}{5}$ | $1\frac{1}{5}$ |
| **9.** | $\frac{3}{6} + \frac{5}{6}$ or $\frac{5}{6} + \frac{3}{6}$ | $1\frac{1}{3}$ | $1\frac{2}{6}$ | $\frac{8}{6}$ | $\frac{8}{12}$ |
| **10.** | $\frac{5}{10} + \frac{3}{10}$ or $\frac{3}{10} + \frac{5}{10}$ | $\frac{4}{5}$ | $\frac{4}{10}$ | $\frac{8}{20}$ | $\frac{8}{10}$ |
| **11.** | $\frac{5}{7} + \frac{6}{7}$ or $\frac{6}{7} + \frac{5}{7}$ | $\frac{11}{14}$ | $\frac{11}{7}$ | $1\frac{5}{7}$ | $1\frac{4}{7}$ |
| **12.** | $\frac{9}{12} + 0$ or $0 + \frac{9}{12}$ | $0$ | $\frac{2}{3}$ | $\frac{3}{4}$ | $\frac{9}{12}$ |
| **13.** | $\frac{5}{8} + \frac{7}{8}$ or $\frac{7}{8} + \frac{5}{8}$ | $\frac{6}{4}$ | $\frac{12}{8}$ | $1\frac{4}{8}$ | $1\frac{1}{2}$ |
| **14.** | $\frac{9}{10} + \frac{5}{10}$ or $\frac{5}{10} + \frac{9}{10}$ | $1\frac{2}{5}$ | $\frac{14}{10}$ | $1\frac{4}{10}$ | $\frac{7}{5}$ |
| **15.** | $\frac{6}{12} + \frac{9}{12}$ or $\frac{9}{12} + \frac{6}{12}$ | $1\frac{3}{12}$ | $1\frac{1}{4}$ | $\frac{15}{12}$ | $\frac{5}{4}$ |


Copyright © William H. Sadlier, Inc. All rights reserved.

# Estimating Sums and Differences

Name ______________________

Date ______________________

**Estimate the sum or difference.**

$\frac{2}{11} + \frac{5}{8}$ → $0 + \frac{1}{2} = \frac{1}{2}$

So $\frac{2}{11} + \frac{5}{8}$ is close to $\frac{1}{2}$.

$\frac{9}{10} - \frac{7}{12}$ → $1 - \frac{1}{2} = \frac{1}{2}$

So $\frac{9}{10} - \frac{7}{12}$ is close to $\frac{1}{2}$.

$\frac{4}{9} + \frac{11}{13}$ → $\frac{1}{2} + 1 = 1\frac{1}{2}$

So $\frac{4}{9} + \frac{11}{13}$ is close to $1\frac{1}{2}$.

**Estimate the sum or difference.**

1. $\frac{1}{4} + \frac{7}{13}$ ______
2. $\frac{6}{13} - \frac{1}{5}$ ______
3. $\frac{5}{9} + \frac{13}{15}$ ______
4. $\frac{8}{9} - \frac{10}{19}$ ______
5. $\frac{9}{11} + \frac{8}{9}$ ______
6. $\frac{7}{8} - \frac{5}{6}$ ______
7. $\frac{1}{7} + \frac{2}{11}$ ______
8. $\frac{8}{17} - \frac{7}{15}$ ______
9. $\frac{9}{10} - \frac{1}{7}$ ______
10. $\frac{10}{19} + \frac{1}{8} + \frac{3}{5}$ ______
11. $\frac{1}{10} + \frac{8}{15} + \frac{4}{9}$ ______
12. $\frac{7}{12} + \frac{2}{9} + \frac{11}{20} + \frac{8}{15}$ ______

13. $10\frac{1}{2} - 8\frac{3}{5}$
14. $5\frac{4}{7} + 6\frac{1}{8}$
15. $12\frac{1}{8} + 11\frac{5}{6}$
16. $15\frac{4}{9} - 3\frac{5}{8}$
17. $9\frac{1}{6} - 8\frac{7}{8}$
18. $11\frac{4}{5} + 2\frac{1}{2}$
19. $13\frac{5}{11} - 1\frac{9}{12}$
20. $6\frac{7}{9} + 14\frac{1}{12}$

**PROBLEM SOLVING Use estimation.**

21. Julio has a 12-inch strip of ribbon. He cuts off a strip that is $6\frac{3}{4}$ inches long. He needs another strip that is $6\frac{3}{4}$ inches long. Does he have enough ribbon? ______________________

22. Joanne set a goal to hike 14 miles in 3 days. She hiked $3\frac{5}{6}$ miles on the first day, $5\frac{7}{9}$ miles on the second day, and $4\frac{2}{3}$ miles on the last day. Did she make her goal of 14 miles? Explain. ______________________


Copyright © William H. Sadlier, Inc. All rights reserved.

# Adding Fractions

Name ______________________

Date ______________________

Add: $\frac{3}{5} + \frac{7}{10} + \frac{1}{2}$

First estimate:

$\frac{1}{2} + \frac{1}{2} + \frac{1}{2} = 1\frac{1}{2}$

Then add:

$$\frac{3}{5} = \frac{3 \times 2}{5 \times 2} = \frac{6}{10}$$
$$\frac{7}{10} = \frac{7}{10}$$
$$+\frac{1}{2} = \frac{1 \times 5}{2 \times 5} = \frac{5}{10}$$
$$\frac{18}{10} = 1\frac{8}{10} = 1\frac{4}{5}$$

The LCD of $\frac{3}{5}$; $\frac{7}{10}$, and $\frac{1}{2}$ is 10.

**Find the LCD for each set of fractions.**

1. $\frac{3}{4}, \frac{5}{16}$ ____
2. $\frac{3}{5}, \frac{1}{9}$ ____
3. $\frac{5}{6}, \frac{7}{9}, \frac{2}{3}$ ____
4. $\frac{3}{8}, \frac{1}{4}, \frac{5}{6}$ ____

**Add. Estimate to help you.**

5. $\frac{1}{8} + \frac{3}{4}$
6. $\frac{4}{7} + \frac{3}{5}$
7. $\frac{1}{6} + \frac{7}{8}$
8. $\frac{2}{3} + \frac{5}{12}$
9. $\frac{5}{14} + \frac{3}{7}$
10. $\frac{9}{24} + \frac{5}{12}$
11. $\frac{5}{6} + \frac{7}{18}$
12. $\frac{4}{5} + \frac{7}{12}$
13. $\frac{7}{12} + \frac{7}{9}$
14. $\frac{8}{9} + \frac{1}{2}$
15. $\frac{5}{9} + \frac{7}{12} + \frac{1}{3}$
16. $\frac{3}{8} + \frac{1}{6} + \frac{3}{4}$
17. $\frac{7}{9} + \frac{1}{2} + \frac{1}{6}$
18. $\frac{3}{4} + \frac{5}{6} + \frac{3}{8}$
19. $\frac{1}{2} + \frac{1}{3} + \frac{1}{4}$

**Compare. Write $<$, $=$, or $>$.**

20. $\frac{1}{4} + \frac{3}{16}$ ____ $\frac{1}{2}$
21. $\frac{1}{8} + \frac{7}{12}$ ____ $\frac{2}{3}$
22. $\frac{2}{5} + \frac{4}{15}$ ____ $\frac{2}{3}$
23. $\frac{2}{3} + \frac{3}{4}$ ____ $1\frac{1}{4}$
24. $\frac{7}{10} + \frac{1}{3}$ ____ $1\frac{1}{30}$
25. $\frac{3}{4} + \frac{3}{8}$ ____ $\frac{7}{8}$

 Copyright © William H. Sadlier, Inc. All rights reserved.

# Adding Mixed Numbers

Name ______________________

Date ______________________

Add: $2\frac{4}{5} + 3\frac{2}{5}$

First estimate: $2\frac{4}{5} + 3\frac{2}{5}$

$3 + 3 = 6$

Then add:

$$\begin{array}{r} 2\frac{4}{5} \\ +\ 3\frac{2}{5} \\ \hline 5\frac{6}{5} = 5 + 1\frac{1}{5} = 6\frac{1}{5} \end{array}$$

Add: $2\frac{4}{10} + 3\frac{2}{5}$

First estimate: $2\frac{4}{10} + 3\frac{2}{5}$

$2 + 3 = 5$

Then add:

$$\begin{array}{r} 2\frac{4}{10} = 2\frac{4}{10} \\ +\ 3\frac{2}{5} = 3\frac{4}{10} \\ \hline 5\frac{8}{10} = 5\frac{4}{5} \end{array}$$

**Rename each in simplest form.**

1. $8\frac{3}{12}$ ______ 2. $7\frac{16}{24}$ ______ 3. $4\frac{6}{22}$ ______ 4. $5\frac{10}{10}$ ______ 5. $16\frac{18}{12}$ ______

**Complete.**

6. $$\begin{array}{r} 9\frac{1}{4} = 9\frac{}{20} \\ +\ 1\frac{3}{5} = 1\frac{}{20} \\ \hline 10\frac{}{20} \end{array}$$

7. $$\begin{array}{r} 12\frac{4}{7} \\ +\ 3\frac{3}{7} \\ \hline 15\frac{}{} = \text{\_\_\_\_} \end{array}$$

8. $$\begin{array}{r} 10\frac{5}{18} = 10\frac{5}{} \\ +\ 11\frac{1}{6} = 11\frac{3}{} \\ \hline 21\frac{}{18} = \text{\_\_\_\_} \end{array}$$

**Add. Estimate to help you.**

9. $4\frac{1}{7} + 5\frac{3}{7}$

10. $1\frac{3}{4} + 3\frac{5}{8}$

11. $4\frac{5}{6} + 6\frac{7}{18}$

12. $7\frac{1}{2} + 9\frac{1}{4}$

13. $8\frac{3}{5} + 7\frac{2}{5}$

14. $12\frac{9}{10} + 1\frac{7}{15}$

15. $1\frac{3}{5} + 3\frac{2}{3}$

16. $2\frac{3}{7} + 4$

17. $3\frac{2}{3} + 6\frac{3}{4} + 2\frac{5}{8} =$ ____

18. $8\frac{1}{2} + 9\frac{3}{4} + 2\frac{2}{5} =$ ____

**PROBLEM SOLVING Express your answer in simplest form.**

19. On a weekend hike, a group walked $5\frac{1}{10}$ miles on Saturday and $4\frac{5}{10}$ miles on Sunday. How many miles did they hike both days? ______________________

20. Malik rode his bicycle $4\frac{1}{2}$ miles in the morning. In the afternoon he rode $3\frac{3}{4}$ miles. How many miles did he ride in all? ______________________


Copyright © William H. Sadlier, Inc. All rights reserved.

# Subtracting Fractions

Name ______________________

Date ______________________

Subtract: $\frac{5}{6} - \frac{2}{8}$

First estimate:

$$\frac{5}{6} - \frac{2}{8}$$
$$\downarrow \quad \downarrow$$
$$1 - 0 = 1$$

Then subtract:

$$\frac{5}{6} = \frac{5 \times 4}{6 \times 4} = \frac{20}{24}$$
$$-\frac{2}{8} = \frac{2 \times 3}{8 \times 3} = \frac{6}{24}$$
$$\frac{14}{24} = \frac{7}{12}$$

The LCD of $\frac{5}{6}$ and $\frac{2}{8}$ is 24.

**Subtract. Estimate to help you. Write the difference in simplest form.**

1. $\frac{5}{8} - \frac{1}{8}$
2. $\frac{2}{3} - \frac{1}{9}$
3. $\frac{3}{4} - \frac{5}{12}$
4. $\frac{3}{7} - \frac{2}{5}$
5. $\frac{1}{2} - \frac{3}{16}$

6. $\frac{3}{4} - \frac{4}{9} =$ ______
7. $\frac{5}{6} - \frac{3}{5} =$ ______
8. $\frac{2}{3} - \frac{1}{4} =$ ______
9. $\frac{5}{6} - \frac{4}{7} =$ ______

10. $\frac{1}{6} - \frac{1}{9}$
11. $\frac{3}{4} - \frac{2}{3}$
12. $\frac{11}{12} - \frac{7}{12}$
13. $\frac{7}{8} - \frac{1}{4}$
14. $\frac{4}{5} - \frac{4}{10}$

**Use a related sentence to find the missing fraction or whole number.**

15. $n + \frac{4}{5} = \frac{7}{8}$ ______
16. $z - \frac{2}{7} = \frac{3}{14}$ ______
17. $\frac{5}{9} = y + \frac{1}{3}$ ______
18. $\frac{1}{4} = x - \frac{1}{16}$ ______
19. $b - 2\frac{1}{3} = 2\frac{2}{3}$ ______
20. $h - 0 = \frac{1}{8}$ ______
21. $t - \frac{8}{9} = 8\frac{1}{9}$ ______
22. $\frac{3}{8} = a + \frac{5}{24}$ ______
23. $\frac{4}{7} = c + \frac{11}{21}$ ______

**PROBLEM SOLVING**

24. Kahn lives $\frac{5}{8}$ mile from school. Frank lives $\frac{3}{5}$ mile from school. How much farther from school does Kahn live? ______________

25. Sue must practice piano $\frac{3}{4}$ h each day. If she has practiced for $\frac{1}{6}$ h, how much longer does she need to practice? ______________


Use with Lesson 6-5, text pages 210–211.
Copyright © William H. Sadlier, Inc. All rights reserved.

# Subtracting Mixed Numbers

Name ____________________

Date ____________________

Subtract: $8\frac{1}{10} - 2\frac{3}{5}$

$$\begin{array}{r} 8\frac{1}{10} \\ -\ 2\frac{3}{5} \\ \hline \end{array} \leftarrow \quad 8\frac{1}{10} = 7 + 1 + \frac{1}{10} = 7 + \frac{10}{10} + \frac{1}{10} = 7\frac{11}{10}$$

$$\begin{array}{r} 7\frac{11}{10} = 7\frac{11}{10} \\ -\ 2\frac{3}{5} = 2\frac{6}{10} \\ \hline 5\frac{5}{10} = 5\frac{1}{2} \end{array}$$

The LCD of $\frac{11}{10}$ and $\frac{3}{5}$ is 10.

**Rename.**

1. $6\frac{1}{4} = 5\frac{\phantom{0}}{4}$
2. $5\frac{2}{3} = 4\frac{\phantom{0}}{3}$
3. $8\frac{3}{7} = 7\frac{\phantom{0}}{7}$
4. $10\frac{1}{8} = 9\frac{\phantom{0}}{8}$
5. $5\frac{2}{5} = 4\frac{\phantom{0}}{5}$
6. $2\frac{2}{3} = 1\frac{\phantom{0}}{3}$
7. $3\frac{5}{7} = 2\frac{\phantom{0}}{7}$
8. $6\frac{4}{5} = 5\frac{\phantom{0}}{5}$
9. $9\frac{3}{4} = 8\frac{\phantom{0}}{4}$
10. $8\frac{5}{6} = 7\frac{\phantom{0}}{6}$

**Subtract. Estimate to help you.**

11. $6\frac{1}{4} - 3\frac{3}{4}$
12. $5\frac{2}{15} - 4\frac{4}{15}$
13. $10\frac{2}{3} - 5\frac{7}{8}$
14. $8\frac{1}{6} - 4\frac{5}{12}$
15. $14\frac{1}{2} - 9\frac{5}{8}$
16. $20\frac{3}{8} - 17\frac{1}{2}$
17. $19 - 12\frac{2}{5}$
18. $15\frac{4}{9} - 12$
19. $6 - 2\frac{7}{8} =$ ______
20. $12\frac{2}{3} - 9 =$ ______
21. $12 - 10\frac{1}{9} =$ ______
22. $8\frac{1}{5} - \frac{2}{3} =$ ______
23. $5\frac{2}{9} - 4\frac{1}{5} =$ ______
24. $10\frac{1}{2} - 6\frac{1}{3} =$ ______

**PROBLEM SOLVING**

25. Larry lives 3 mi from the library. Jalen lives $1\frac{5}{8}$ mi from the library. How much farther from the library does Larry live than Jalen? ______

26. Jane baked $8\frac{1}{4}$ dozen cupcakes. She sold all but $\frac{5}{6}$ dozen. How many dozen cupcakes did she sell? ______


Copyright © William H. Sadlier, Inc. All rights reserved.

# Properties and Mixed Numbers

Name ______________________

Date ______________________

**Compute:** $14\frac{3}{7} + 5\frac{2}{3} - 14\frac{3}{7}$

$5\frac{2}{3} + 14\frac{3}{7} - 14\frac{3}{7}$ ← **Commutative Property**

$5\frac{2}{3} + (14\frac{3}{7} - 14\frac{3}{7})$ ← **Associative Property**

$5\frac{2}{3} + 0 = 5\frac{2}{3}$ ← **Identity Property**

**Name the property of addition.**

1. $4\frac{7}{8} + 0 = 4\frac{7}{8}$ ______________
2. $7\frac{2}{3} + 1\frac{5}{6} = 1\frac{5}{6} + 7\frac{2}{3}$ ______________
3. $(1\frac{1}{5} + 2\frac{1}{3}) + 1\frac{2}{3} = 1\frac{1}{5} + (2\frac{1}{3} + 1\frac{2}{3})$ ______________
4. $0 + 6\frac{2}{3} = 6\frac{2}{3} + 0$ ______________
5. $0 + 12\frac{1}{8} = 12\frac{1}{8}$ ______________
6. $7\frac{3}{4} + (1\frac{1}{4} + 3\frac{1}{2}) = (7\frac{3}{4} + 1\frac{1}{4}) + 3\frac{1}{2}$ ______________

**Add. Use mental math and the properties of addition**

7. $\frac{3}{4} + \frac{1}{2} =$ ______
8. $\frac{1}{4} + \frac{1}{8} =$ ______
9. $\frac{1}{6} + \frac{1}{6} =$ ______
10. $\frac{1}{3} + \frac{1}{6} =$ ______
11. $2\frac{3}{4} + 4\frac{1}{2} =$ ______
12. $1\frac{1}{4} + 2\frac{3}{8} =$ ______
13. $3\frac{1}{3} + 5\frac{1}{6} =$ ______
14. $1\frac{2}{3} + 4\frac{1}{6} =$ ______
15. $\frac{1}{2} + \frac{1}{6} =$ ______
16. $\frac{1}{2} + \frac{5}{6} =$ ______
17. $4\frac{1}{8} + 3\frac{1}{2} =$ ______
18. $9\frac{1}{4} + 2\frac{3}{8} =$ ______

**Compute. Name the properties of addition you used.**

19. $2\frac{5}{7} + 3\frac{1}{7} + 6\frac{1}{4} =$ ______ ______________
20. $12\frac{2}{3} + 0 =$ ______ ______________
21. $3\frac{1}{4} + 6\frac{1}{2} + 2 =$ ______ ______________
22. $1\frac{5}{9} + 3\frac{2}{3} + 6\frac{1}{9} =$ ______ ______________
23. $2\frac{1}{8} + 3\frac{1}{4} + 1\frac{3}{8} =$ ______ ______________
24. $1\frac{1}{3} + (4\frac{5}{6} + 2\frac{1}{3}) =$ ______ ______________
25. $0 + 15\frac{7}{8} =$ ______ ______________
26. $8\frac{2}{3} + (2\frac{3}{5} - 2\frac{3}{5}) =$ ______ ______________
27. $(8\frac{1}{6} + 1\frac{2}{3}) + (7\frac{4}{9} - 7\frac{4}{9}) =$ ______ ______________

**PROBLEM SOLVING**

28. Su Lin bought $4\frac{3}{8}$ yards of blue fabric, $3\frac{1}{2}$ yards of red fabric, and $5\frac{1}{8}$ yards of green fabric. How many yards of fabric did she buy? ______________


Copyright © William H. Sadlier, Inc. All rights reserved.

# Problem-Solving Strategy: Working Backwards

Name ____________________

Date ____________________

Aunt Martha says, "If you subtract 8 from my age and divide the result by 2, you will get the number of eggs in a dozen." How old is Aunt Martha?

| | |
|---|---|
| Start with the number of eggs in a dozen. | 12 |
| Multiply by 2. | $2 \times 12 = 24$ |
| Add 8. | $24 + 8 = 32$ |

Aunt Martha is 32 years old.

**PROBLEM SOLVING Do your work on a separate sheet of paper.**

1. Shawna had to decorate 3 gift packages with ribbon. She used $1\frac{1}{3}$ yards of ribbon for the first package, $1\frac{1}{2}$ yards of ribbon for the second package, and $1\frac{2}{3}$ yards to decorate the third package. Shawna had $3\frac{1}{2}$ yards of ribbon left over. How much ribbon did she start with?

   ____________________

2. Paulina gets on an elevator. At the first stop, 4 people get off and 3 get on. At the second stop, 5 people get off and one gets on. At the third stop, Paulina gets off. If 4 people are still in the elevator when Paulina gets off, how many people were on the elevator when she got on?

   ____________________

3. From one piece of walnut, Sam cut two $9\frac{1}{2}$ in. strips for 2 sides of a frame. He cut two $11\frac{3}{4}$ in. strips each for the other 2 sides. Sam had $5\frac{1}{2}$ in. of walnut left. How long was the piece of walnut when he started?

   ____________________

4. Jeanine used $2\frac{1}{2}$ cups of flour to make muffins and $1\frac{3}{4}$ cups of flour to make biscuits. After borrowing $\frac{1}{4}$ cup of flour, she had $2\frac{1}{4}$ cups left to make bread. How much flour did she have to begin with?

   ____________________

5. Simon added $4.50 to the money in his bank. Then his father agreed to double Simon's money, which gave him $33. How much money did Simon have to begin with?

   ____________________

6. Franklin was paid for doing yard work. He spent $4.45 of his pay for a sandwich and fruit drink and $2.95 for a magazine. He had $8.55 left. How much was Franklin paid for the yard work?

   ____________________

7. Chita is $3\frac{3}{4}$ years younger than Pablo. If you multiply Chita's age by 4, the answer is 43. How old is Pablo?

   ____________________

8. Duwayne wrote a two-digit number. He multiplied it by 3, added 18, and divided by 8. His final answer was 9. What number did Duwayne write?

   ____________________


Copyright © William H. Sadlier, Inc. All rights reserved.

# Multiplying Fractions by Fractions

Name ______________________

Date ______________________

**Multiply:** $\frac{2}{3} \times \frac{4}{5}$

$$\frac{2}{3} \times \frac{4}{5} = \frac{2 \times 4}{3 \times 5} = \frac{8}{15}$$

**Multiply:** $\frac{3}{4} \times \frac{8}{15}$

$$\frac{3}{4} \times \frac{8}{15} = \frac{\overset{1}{\cancel{3}}}{\underset{1}{\cancel{4}}} \times \frac{\overset{2}{\cancel{8}}}{\underset{5}{\cancel{15}}} = \frac{1 \times 2}{1 \times 5} = \frac{2}{5}$$

**Complete.**

1. $\frac{3}{4} \times \frac{1}{2} = \frac{3 \times \_}{\_ \times 2} = \frac{\_}{\_}$

2. $\frac{3}{5} \times \frac{10}{21} = \frac{1}{\_} \times \frac{\_}{7} = \frac{\_}{\_}$

3. $\frac{4}{9} \times \frac{3}{8} = \frac{\overset{2}{\cancel{4}}}{\underset{3}{\cancel{9}}} \times \frac{\overset{\_}{\cancel{3}}}{\underset{\_}{\cancel{8}}} = \frac{2}{\_} = \frac{\_}{\_}$

**Multiply. Cancel whenever possible.**

4. $\frac{2}{7} \times \frac{3}{7} =$ ______
5. $\frac{3}{8} \times \frac{1}{4} =$ ______
6. $\frac{1}{3} \times \frac{2}{5} =$ ______
7. $\frac{7}{12} \times \frac{4}{5} =$ ______
8. $\frac{6}{10} \times \frac{4}{5} =$ ______
9. $\frac{12}{15} \times \frac{3}{4} =$ ______
10. $\frac{3}{4} \times \frac{6}{8} =$ ______
11. $\frac{18}{20} \times \frac{5}{10} =$ ______
12. $\frac{12}{14} \times \frac{2}{7} =$ ______
13. $\frac{8}{21} \times \frac{3}{12} =$ ______
14. $\frac{22}{70} \times \frac{10}{11} =$ ______
15. $\frac{7}{16} \times \frac{12}{21} =$ ______
16. $\frac{2}{10} \times \frac{3}{7} \times \frac{5}{9} =$ ______
17. $\frac{3}{4} \times \frac{8}{12} \times \frac{6}{9} =$ ______
18. $\frac{5}{6} \times \frac{1}{10} \times \frac{6}{9} =$ ______

**PROBLEM SOLVING**

19. Philip had $\frac{5}{8}$ yd of paper toweling. He used $\frac{2}{3}$ of it. How much toweling did Phillip use? ______

20. Janelle purchased a piece of wood that measured $\frac{5}{6}$ yd. She used only $\frac{3}{5}$ of it. How much of the wood was unused? ______

21. A punch recipe calls for $\frac{3}{4}$ cup of grape juice. If Carlos wants to make $\frac{1}{2}$ the recipe, how much grape juice should he use? ______


Copyright © William H. Sadlier, Inc. All rights reserved.

# Multiplying Fractions and Whole Numbers

Name ______________________

Date ______________________

Multiply: $5 \times \frac{3}{8}$

$5 \times \frac{3}{8} = \frac{5}{1} \times \frac{3}{8}$

$= \frac{5 \times 3}{1 \times 8}$

$= \frac{15}{8} = 1\frac{7}{8}$

Find: $\frac{2}{3}$ of \$84 — "of" means "times"

$\frac{2}{3} \times 84 = \frac{2}{\cancel{3}_1} \times \frac{\cancel{84}^{28}}{1}$

$= \frac{2 \times 28}{1 \times 1} = \frac{56}{1} = 56$

**Multiply.**

1. $3 \times \frac{2}{3} =$ _____
2. $9 \times \frac{5}{6} =$ _____
3. $14 \times \frac{2}{7} =$ _____
4. $8 \times \frac{1}{7} =$ _____
5. $24 \times \frac{5}{8} =$ _____
6. $15 \times \frac{3}{5} =$ _____
7. $18 \times \frac{3}{4} =$ _____
8. $25 \times \frac{4}{15} =$ _____
9. $\frac{7}{8} \times 56 =$ _____
10. $\frac{4}{9} \times 18 =$ _____
11. $\frac{3}{4} \times 10 =$ _____
12. $\frac{4}{5} \times 8 =$ _____

**Find the product.**

13. $\frac{1}{8}$ of 32 = _____
14. $\frac{1}{5}$ of 25 = _____
15. $\frac{1}{8}$ of 12 = _____
16. $\frac{2}{3}$ of 18 = _____
17. $\frac{2}{9}$ of 3 = _____
18. $\frac{3}{10}$ of 5 = _____
19. $\frac{1}{8}$ of 4 = _____
20. $\frac{2}{3}$ of 14 = _____
21. $\frac{3}{4}$ of \$24 = _____
22. $\frac{2}{3}$ of \$27 = _____
23. $\frac{1}{5}$ of \$3.50 = _____
24. $\frac{3}{7}$ of \$4.20 = _____

**PROBLEM SOLVING**

25. David lives $\frac{7}{8}$ mi from school. If he walks to school but does not walk home each day, how many miles does he walk in a 5-day week? ______________

26. Susan baked 12 muffins. Her family ate $\frac{5}{6}$ of them. How many muffins were left? ______________

27. Joshua rode his bicycle 16 miles. He stopped for a rest after riding $\frac{3}{4}$ of the way. How many miles did he ride after resting? ______________


Copyright © William H. Sadlier, Inc. All rights reserved.

# Properties and the Reciprocal

Name ______________________

Date ______________________

| Commutative Property | Associative Property | Identity Property |
| --- | --- | --- |
| $\frac{3}{8} \times \frac{4}{7} = \frac{4}{7} \times \frac{3}{8}$ | $(\frac{1}{3} \times \frac{1}{4}) \times 7 = \frac{1}{3} \times (\frac{1}{4} \times 7)$ | $1 \times \frac{2}{9} = \frac{2}{9}$<br>$\frac{2}{9} \times 1 = \frac{2}{9}$ |
| **Zero Property** | **Reciprocals** | |
| $0 \times \frac{8}{9} = 0$<br>$\frac{8}{9} \times 0 = 0$ | $\frac{2}{3} \times \frac{3}{2} = \frac{\overset{1}{\cancel{2}}}{\underset{1}{\cancel{3}}} \times \frac{\overset{1}{\cancel{3}}}{\underset{1}{\cancel{2}}} = 1$ | $\frac{2}{3}$ and $\frac{3}{2}$ are reciprocals. |

**Complete. Use the properties of multiplication.**

1. $\frac{1}{8} \times \frac{2}{5} = \frac{2}{5} \times$ ____
2. $\frac{5}{12} \times 0 =$ ____
3. $\frac{4}{5} \times$ ____ $= \frac{4}{5}$
4. ____ $\times \frac{3}{10} = 0$
5. $\frac{1}{9} \times \frac{2}{3} =$ ____ $\times \frac{1}{9}$
6. ____ $\times \frac{3}{7} = \frac{3}{7}$
7. $(\frac{2}{5} \times \frac{1}{11}) \times \frac{1}{6} = \frac{2}{5} \times ($ ____ $\times \frac{1}{6})$
8. $\frac{4}{5} \times (12 \times$ ____ $) = (\frac{4}{5} \times 12) \times \frac{1}{6}$

**Write the missing reciprocal in each number sentence.**

9. $6 \times$ ____ $= 1$
10. $\frac{4}{5} \times$ ____ $= 1$
11. $8 \times$ ____ $= 1$
12. $\frac{1}{13} \times$ ____ $= 1$
13. $55 \times$ ____ $= 1$
14. $200 \times$ ____ $= 1$
15. $\frac{8}{7} \times$ ____ $= 1$
16. $\frac{2}{5} \times$ ____ $= 1$

**Write the reciprocal of each number.**

17. 3 ____
18. 23 ____
19. $\frac{1}{4}$ ____
20. $\frac{13}{24}$ ____
21. $\frac{9}{7}$ ____
22. $\frac{15}{8}$ ____

**Compute. Name the properties used.**

23. $\frac{4}{7} \times \frac{2}{3} \times 6 =$ ______
24. $0 \times \frac{7}{12} \times \frac{10}{11} =$ ______
25. $(\frac{1}{5} \times \frac{3}{8}) \times 15 =$ ______
26. $(\frac{2}{5} \times \frac{1}{3}) \times 3 =$ ______
27. $\frac{4}{3} \times (\frac{3}{4} \times \frac{15}{28}) =$ ______
28. $(\frac{5}{6} \times \frac{2}{3}) \times \frac{6}{5} =$ ______
29. $\frac{2}{5} \times \frac{1}{6} \times \frac{5}{8} =$ ______
30. $\frac{4}{9} \times 8 \times 18 =$ ______
31. $\frac{5}{24} \times \frac{12}{25} \times \frac{24}{5} =$ ______

 Use with Lesson 7-3, text pages 230–231. Copyright © William H. Sadlier, Inc. All rights reserved.

# Multiplying Mixed Numbers

Name ______________________

Date ______________________

Estimate: $4\frac{2}{3} \times 14\frac{1}{4}$

$4\frac{2}{3} \times 14\frac{1}{4} \longrightarrow 5 \times 14 = 70$

So $4\frac{2}{3} \times 14\frac{1}{4}$ is about 70.

Multiply: $4\frac{2}{3} \times 14\frac{1}{4}$

$4\frac{2}{3} \times 14\frac{1}{4} = \frac{14}{3} \times \frac{57}{4}$

$= \frac{\overset{7}{\cancel{14}}}{\underset{1}{\cancel{3}}} \times \frac{\overset{19}{\cancel{57}}}{\underset{2}{\cancel{4}}}$

$= \frac{133}{2}$

$= 66\frac{1}{2}$ ← $66\frac{1}{2}$ is close to the estimate of 70.

**Estimate each product.**

1. $3\frac{5}{6} \times 8\frac{2}{7} =$ ____
2. $6\frac{1}{5} \times 4\frac{3}{4} =$ ____
3. $7 \times 9\frac{7}{8} =$ ____
4. $93\frac{5}{6} \times \frac{4}{7} =$ ____
5. $15\frac{1}{3} \times 2\frac{1}{4} =$ ____
6. $\frac{4}{5} \times 28\frac{2}{3} =$ ____

**Multiply. Estimate to help you.**

7. $4\frac{2}{3} \times 18 =$ ____
8. $1\frac{1}{6} \times 4\frac{4}{5} =$ ____
9. $2\frac{3}{4} \times 3\frac{1}{5} =$ ____
10. $\frac{3}{7} \times 8\frac{1}{6} =$ ____
11. $5\frac{1}{3} \times 3\frac{3}{8} =$ ____
12. $1\frac{1}{8} \times 10\frac{2}{3} =$ ____
13. $21 \times 3\frac{2}{3} =$ ____
14. $1\frac{1}{5} \times 5\frac{2}{3} \times 2\frac{1}{2} =$ ____
15. $1\frac{4}{5} \times 2\frac{1}{3} \times 1\frac{7}{8} =$ ____

**Compare. Write <, =, or >.**

16. $4\frac{1}{3} \times 5\frac{2}{7} \times 2\frac{1}{2}$ ____ $4\frac{2}{7} \times 5\frac{1}{3}$
17. $1\frac{1}{2} \times 3\frac{1}{4}$ ____ $1\frac{1}{4} \times 3\frac{1}{2}$
18. $2\frac{3}{5} \times 1\frac{1}{5}$ ____ $1\frac{4}{5} \times 1\frac{2}{3}$
19. $2\frac{4}{9} \times 1\frac{1}{2}$ ____ $2\frac{2}{3} \times 1\frac{1}{3}$

**Find the value of *n*. Use the properties of multiplication.**

20. $n \times 4\frac{2}{5} = 0$ ____
21. $1\frac{1}{2} \times n = \frac{3}{4} \times 1\frac{1}{2}$ ____
22. $1 \times n = 7\frac{4}{5}$ ____
23. $(\frac{2}{3} \times \frac{1}{4}) \times 4\frac{1}{3} = \frac{2}{3} \times (n \times 4\frac{1}{3})$ ____
24. $\frac{7}{8} \times (3\frac{1}{2} \times n) = (\frac{7}{8} \times 4) \times 3\frac{1}{2}$ ____

**PROBLEM SOLVING**

25. A group hiked $2\frac{3}{8}$ miles in one hour. At that rate, how far could it hike in $2\frac{1}{2}$ hours? ______________

26. Sean is $15\frac{1}{2}$ years old. Kathy is $1\frac{1}{3}$ times as old. How old is Kathy? ______________


Copyright © William H. Sadlier, Inc. All rights reserved.

# Meaning of Division

Name ______________________________

Date ______________________________

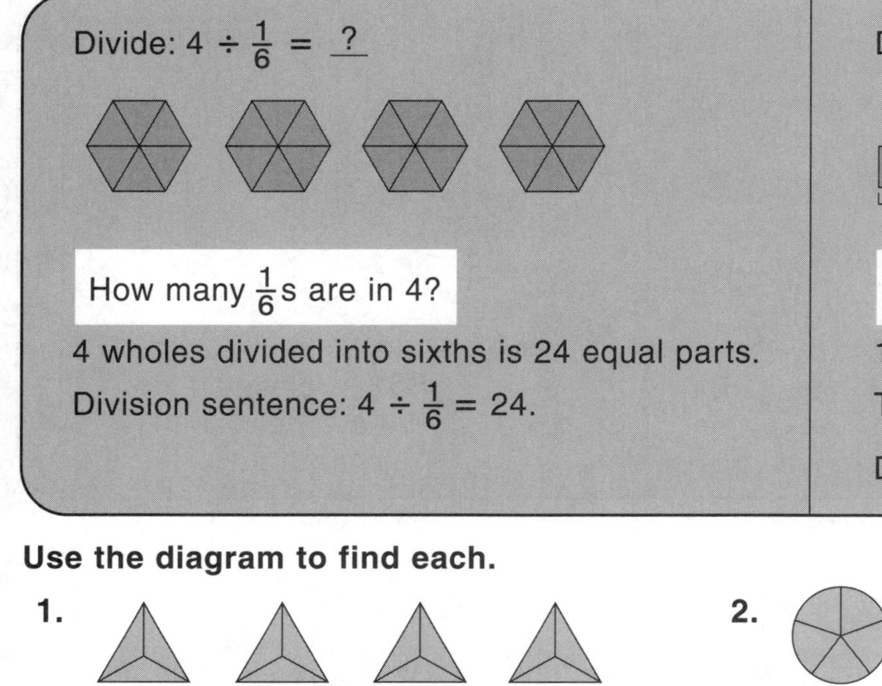

Divide: $4 \div \frac{1}{6} = \underline{?}$

How many $\frac{1}{6}$s are in 4?

4 wholes divided into sixths is 24 equal parts.

Division sentence: $4 \div \frac{1}{6} = 24$.

Divide: $\frac{4}{5} \div \frac{2}{5} = \underline{?}$

How many $\frac{2}{5}$s are in $\frac{4}{5}$?

1 whole is divided into fifths.

There are two $\frac{2}{5}$'s in $\frac{4}{5}$.

Division sentence: $\frac{4}{5} \div \frac{2}{5} = 2$.

**Use the diagram to find each.**

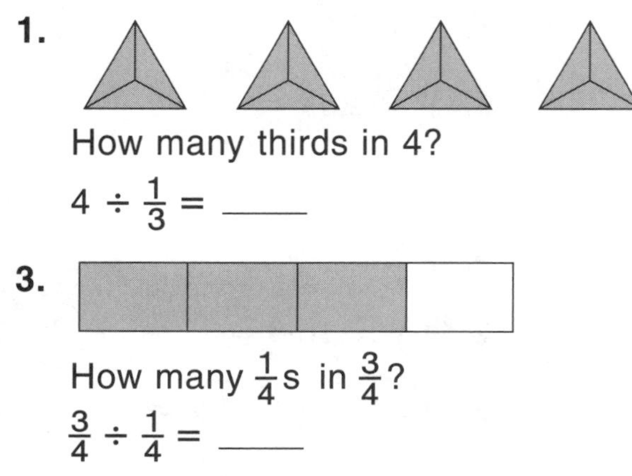

**1.** How many thirds in 4?

$4 \div \frac{1}{3} =$ _____

**2.** How many fifths in 3?

$3 \div \frac{1}{5} =$ _____

**3.** How many $\frac{1}{4}$s in $\frac{3}{4}$?

$\frac{3}{4} \div \frac{1}{4} =$ _____

**4.** How many $\frac{3}{10}$s in $\frac{9}{10}$?

$\frac{9}{10} \div \frac{3}{10} =$ _____

**Write a division sentence for each diagram.**

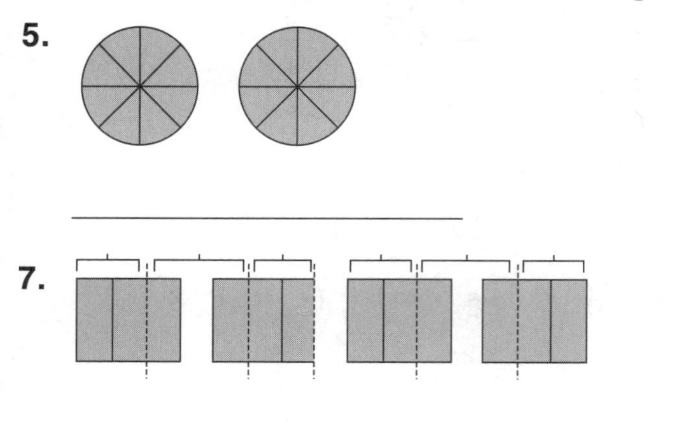

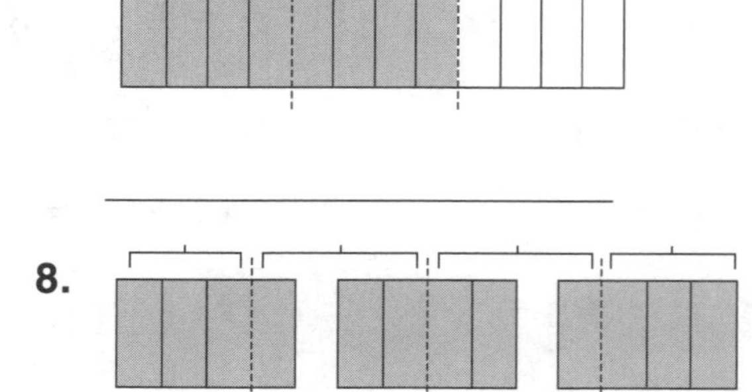

**5.** ______________________________

**6.** ______________________________

**7.** ______________________________

**8.** ______________________________

**Draw a diagram to show each division.**

**9.** $5 \div \frac{1}{2}$

**10.** $\frac{4}{6} \div \frac{1}{6}$


Copyright © William H. Sadlier, Inc. All rights reserved.

# Dividing Fractions by Fractions

Name ______________________

Date ______________________

Divide: $\frac{3}{4} \div \frac{5}{8}$

$\frac{3}{4} \div \frac{5}{8} = \frac{3}{\cancel{4}_1} \times \frac{\cancel{8}^2}{5} = \frac{6}{5} = 1\frac{1}{5}$

reciprocals

Divide: $\frac{2}{3} \div \frac{1}{6}$

$\frac{2}{3} \div \frac{1}{6} = \frac{2}{\cancel{3}_1} \times \frac{\cancel{6}^2}{1} = 4$

reciprocals

**Complete.**

1. $\frac{4}{7} \div \frac{2}{3} = \frac{4}{7} \times \frac{3}{2} =$ _____
2. $\frac{8}{15} \div \frac{16}{25} = \frac{8}{15} \times \frac{25}{16} =$ _____
3. $\frac{5}{6} \div \frac{5}{24} = \frac{5}{6} \times \frac{24}{5} =$ _____
4. $\frac{3}{4} \div \frac{1}{16} = \frac{3}{4} \times \frac{16}{1} =$ _____

**Divide.**

5. $\frac{1}{2} \div \frac{1}{8} =$ _____
6. $\frac{4}{5} \div \frac{8}{15} =$ _____
7. $\frac{9}{16} \div \frac{3}{4} =$ _____
8. $\frac{7}{12} \div \frac{7}{9} =$ _____
9. $\frac{1}{8} \div \frac{1}{7} =$ _____
10. $\frac{20}{21} \div \frac{5}{14} =$ _____
11. $\frac{7}{8} \div \frac{21}{32} =$ _____
12. $\frac{16}{21} \div \frac{4}{7} =$ _____
13. $\frac{8}{9} \div \frac{14}{15} =$ _____

**PROBLEM SOLVING**

14. How many $\frac{1}{8}$s are there in $\frac{3}{4}$? __________
15. How many $\frac{1}{20}$s are there in $\frac{1}{5}$? __________
16. Marty had $\frac{7}{8}$ of a bottle of milk. He gave each of his friends $\frac{1}{8}$ of the milk. How many friends did he share with? __________


Copyright © William H. Sadlier, Inc. All rights reserved.

# Estimation in Division

Name ______________________

Date ______________________

| When the dividend is greater than the divisor, the quotient is greater than 1. | When the dividend is less than the divisor, the quotient is less than 1. | Round mixed numbers to whole numbers to estimate quotients. |
|---|---|---|
| $\frac{3}{7} \div \frac{1}{7} = 3$ <br> $\frac{3}{7} > \frac{1}{7}$ $3 > 1$ | $\frac{1}{6} \div \frac{5}{6} = \frac{1}{5}$ <br> $\frac{1}{6} < \frac{5}{6}$ $\frac{1}{5} < 1$ | $4\frac{2}{3} \div 14\frac{1}{2} = \underline{?}$ <br> $4\frac{2}{3} < 14\frac{1}{2}$, so the quotient is less than 1. <br> $4\frac{2}{3} \div 14\frac{1}{2}$ <br> $\downarrow \quad \downarrow$ <br> $5 \div 15 = \frac{1}{3}$ ← estimated quotient |

**Write whether the quotient is less than 1 or greater than 1.**

1. $\frac{1}{3} \div \frac{2}{3}$ ______
2. $\frac{5}{9} \div \frac{2}{9}$ ______
3. $\frac{1}{4} \div \frac{1}{12}$ ______
4. $\frac{1}{11} \div \frac{1}{10}$ ______
5. $\frac{2}{7} \div \frac{2}{5}$ ______
6. $\frac{5}{8} \div \frac{5}{12}$ ______
7. $\frac{3}{4} \div \frac{1}{3}$ ______
8. $\frac{3}{4} \div \frac{7}{8}$ ______
9. $\frac{1}{3} \div \frac{5}{12}$ ______
10. $\frac{13}{24} \div \frac{7}{12}$ ______
11. $\frac{3}{10} \div \frac{7}{15}$ ______
12. $\frac{3}{8} \div \frac{6}{7}$ ______
13. $1 \div \frac{1}{4}$ ______
14. $1\frac{1}{5} \div 1\frac{1}{2}$ ______
15. $3\frac{2}{3} \div \frac{7}{8}$ ______
16. $2\frac{4}{5} \div 1\frac{2}{7}$ ______

**Estimate. Round each mixed number to the nearest whole number.**

17. $3\frac{3}{5} \div 1\frac{7}{8}$ ______
18. $9\frac{1}{4} \div 2\frac{3}{4}$ ______
19. $7 \div 2\frac{1}{3}$ ______
20. $15 \div 1\frac{9}{10}$ ______
21. $3\frac{4}{5} \div 16\frac{1}{7}$ ______
22. $8\frac{2}{7} \div 11\frac{2}{3}$ ______
23. $\frac{5}{6} \div 3\frac{1}{12}$ ______
24. $9\frac{1}{8} \div \frac{9}{16}$ ______

**Compare. Write < or >. Use estimation to help you.**

25. $1 \div \frac{2}{9}$ ______ $1$
26. $\frac{1}{3} \div \frac{9}{10}$ ______ $1$
27. $3 \div \frac{1}{8}$ ______ $\frac{1}{8} \div 3$
28. $12\frac{1}{7} \div 2\frac{1}{3}$ ______ $1$
29. $32 \div 8$ ______ $32 \div 7\frac{4}{9}$
30. $\frac{3}{5} \div 9\frac{1}{4}$ ______ $9\frac{1}{4} \div \frac{3}{5}$

**PROBLEM SOLVING**

31. About how much will each person get if 8 people share $1\frac{1}{3}$ qt of lemonade?

______________________

 **Use with Lesson 7-7, text pages 238–239.** Copyright © William H. Sadlier, Inc. All rights reserved.

# Dividing Whole Numbers

Name ______________________________

Date ______________________________

Divide: $9 \div \frac{4}{7}$

$9 \div \frac{4}{7} = \frac{9}{1} \div \frac{4}{7}$

$= \frac{9}{1} \times \frac{7}{4}$

$= \frac{63}{4} = 15\frac{3}{4}$ ← simplest form

Divide: $\frac{1}{4} \div 8$

$\frac{1}{4} \div 8 = \frac{1}{4} \div \frac{8}{1}$

$= \frac{1}{4} \times \frac{1}{8}$

$= \frac{1}{32}$

**Complete.**

1. $7 \div \frac{1}{3} = \frac{7}{1} \div$ ______
   $= \frac{7}{1} \times$ ______

2. $\frac{1}{9} \div 5 = \frac{1}{9} \div$ ______
   $= \frac{1}{9} \times$ ______

**Divide. Estimate to help you.**

3. $6 \div \frac{1}{4} =$ ________
4. $8 \div \frac{1}{8} =$ ________
5. $3 \div \frac{2}{7} =$ ________
6. $4 \div \frac{2}{5} =$ ________
7. $\frac{1}{3} \div 2 =$ ________
8. $24 \div \frac{8}{9} =$ ________
9. $\frac{4}{5} \div 8 =$ ________
10. $\frac{3}{7} \div 9 =$ ________
11. $6 \div \frac{2}{5} =$ ________
12. $\frac{1}{6} \div 3 =$ ________
13. $28 \div \frac{7}{8} =$ ________
14. $15 \div \frac{5}{6} =$ ________
15. $3 \div \frac{3}{4} =$ ________
16. $\frac{1}{4} \div 4 =$ ________
17. $\frac{4}{5} \div 16 =$ ________

**Compare. Write $<$, $=$, or $>$.**

18. $15 \div \frac{5}{7}$ ____ $\frac{5}{7} \div 15$
19. $\frac{1}{6} \div 8$ ____ $\frac{1}{8} \div 6$
20. $\frac{1}{4} \div 7$ ____ $7 \div \frac{1}{4}$
21. $4 \div \frac{2}{3}$ ____ $6 \div \frac{2}{3}$
22. $\frac{1}{8} \div 3$ ____ $\frac{1}{12} \div 2$
23. $7 \div \frac{12}{19}$ ____ $7 \div \frac{19}{12}$

**PROBLEM SOLVING**

24. How many $\frac{3}{4}$-inch pieces of ribbon can be cut from a ribbon that is 12 inches long? ______________________

25. One batch of honey-nut muffins requires $\frac{2}{3}$ cup of honey. How many batches can Mindy make with 3 cups of honey? ______________________


Copyright © William H. Sadlier, Inc. All rights reserved.

# Dividing a Mixed Number

Name ______________________

Date ______________________

First estimate: $7\frac{7}{8} \div 3\frac{1}{2} = \underline{\ ?\ }$

$7\frac{7}{8} \div 3\frac{1}{2} \longrightarrow 8 \div 4 = 2$

Estimated quotient: 2

Then divide: $7\frac{7}{8} \div 3\frac{1}{2} = \underline{\ ?\ }$

$$7\frac{7}{8} \div 3\frac{1}{2} = \frac{63}{8} \div \frac{7}{2}$$
$$= \frac{63}{8} \times \frac{2}{7}$$
$$= \frac{\overset{9}{\cancel{63}}}{\underset{4}{\cancel{8}}} \times \frac{\overset{1}{\cancel{2}}}{\underset{1}{\cancel{7}}}$$
$$= \frac{9}{4} = 2\frac{1}{4}$$

$2\frac{1}{4}$ is close to the estimate of 2.

**Complete.**

1. $9\frac{1}{3} \div 1\frac{1}{3} = \frac{28}{3} \div$ ______
   $= \frac{28}{3} \times$ ______

2. $5\frac{1}{4} \div 7 = \frac{\ \ }{4} \div$ ______
   $= \frac{\ \ }{4} \times$ ______

**Divide.** Estimate to help you.

3. $8\frac{2}{3} \div 1\frac{1}{3} =$ ______
4. $8\frac{1}{3} \div 1\frac{1}{4} =$ ______
5. $7\frac{1}{5} \div 1\frac{1}{3} =$ ______
6. $\frac{3}{4} \div 4\frac{1}{5} =$ ______
7. $6\frac{2}{3} \div \frac{5}{8} =$ ______
8. $2\frac{1}{7} \div \frac{5}{14} =$ ______
9. $2\frac{2}{5} \div 4 =$ ______
10. $2 \div 1\frac{1}{3} =$ ______
11. $6\frac{1}{9} \div \frac{5}{6} =$ ______
12. $\frac{9}{11} \div 3 =$ ______
13. $5\frac{5}{6} \div \frac{15}{16} =$ ______
14. $6\frac{2}{3} \div 2\frac{2}{5} =$ ______
15. $63 \div 2\frac{5}{8} =$ ______
16. $5\frac{3}{5} \div 1\frac{3}{4} =$ ______
17. $2\frac{7}{9} \div 6\frac{2}{3} =$ ______
18. $9\frac{1}{3} \div \frac{7}{12} =$ ______
19. $5\frac{1}{7} \div 2\frac{2}{7} =$ ______
20. $6\frac{1}{4} \div 1\frac{3}{4} =$ ______

**PROBLEM SOLVING**

21. How many pieces of yarn $\frac{3}{4}$ yd long can be cut from a $10\frac{1}{2}$-yd length of yarn? ______

22. Allen drove 90 miles in $2\frac{1}{4}$ hours. If he drove at a constant rate of speed, how many miles did he drive in one hour? ______


Copyright © William H. Sadlier, Inc. All rights reserved.

# Order of Operations Using Fractions

Name ______________________

Date ______________________

Compute: $2\frac{1}{4} + 4\frac{1}{2} \times \frac{1}{6} = \underline{?}$

$$2\frac{1}{4} + 4\frac{1}{2} \times \frac{1}{6} = \frac{9}{4} + \frac{9}{2} \times \frac{1}{6}$$

$$= \frac{9}{4} + \frac{\overset{3}{\cancel{9}}}{2} \times \frac{1}{\underset{2}{\cancel{6}}}$$ Multiply first.

$$= \frac{9}{4} + \frac{3}{4}$$ Then add.

$$= \frac{12}{4} = 3$$

So $2\frac{1}{4} + 4\frac{1}{2} \times \frac{1}{6} = 3$.

Order of Operations
( ) first
× or ÷ left to right
+ or − left to right

**Compute.**

1. $\frac{4}{7} \div 1\frac{1}{7} + 5 =$ ____
2. $1\frac{3}{4} - \frac{1}{4} - 1 =$ ____
3. $2\frac{1}{3} + 1\frac{1}{3} - 2 =$ ____
4. $1 - \frac{1}{8} + 1\frac{5}{8} =$ ____
5. $2\frac{1}{6} + 1\frac{5}{6} - 2\frac{1}{3} =$ ____
6. $\frac{4}{5} - \frac{1}{5} + 8 =$ ____
7. $\frac{1}{3} \times \frac{2}{3} + 6 =$ ____
8. $4 \times \frac{5}{8} \div \frac{1}{3} =$ ____
9. $8 \times \frac{1}{4} \div \frac{1}{2} =$ ____
10. $\frac{3}{8} \div \frac{1}{2} + 1\frac{1}{4} =$ ____
11. $6\frac{1}{3} - 4 \div \frac{3}{4} =$ ____
12. $1\frac{3}{4} \times \frac{1}{7} \div \frac{2}{3} =$ ____
13. $\frac{3}{5} + \frac{4}{5} \times \frac{1}{2} =$ ____
14. $10 \times \frac{2}{5} \div \frac{3}{5} =$ ____
15. $2\frac{1}{5} \times \frac{1}{4} \div \frac{5}{8} =$ ____
16. $\frac{1}{2} \times \frac{2}{3} - \frac{1}{6} =$ ____
17. $\frac{5}{9} - \frac{1}{3} \div 3 =$ ____
18. $\frac{2}{7} + \frac{1}{3} \times 12 =$ ____
19. $(1\frac{1}{3} \times 6) - \frac{2}{3} =$ ____
20. $(1\frac{1}{4} + \frac{3}{4}) \times \frac{5}{8} =$ ____
21. $2\frac{4}{5} \times (\frac{1}{7} + 2\frac{5}{7}) =$ ____
22. $(18 \div 1\frac{1}{5}) \times \frac{1}{10} =$ ____
23. $\frac{2}{5} \times (1\frac{1}{2} - \frac{1}{4}) =$ ____
24. $1\frac{3}{8} \div (\frac{1}{3} + \frac{1}{6}) =$ ____


Copyright © William H. Sadlier, Inc. All rights reserved.

# Fractions with Money

Name ______________________

Date ______________________

| Find: $\frac{2}{5}$ of \$1.40 | Divide: \$4.60 ÷ $1\frac{1}{3}$ |
|---|---|
| $\frac{2}{5}$ of \$1.40 $= \frac{2}{5} \times \$1.40$ <br> $= \frac{2}{5} \times \frac{\$.28}{1}$ <br> $= \frac{\$.56}{1} = \$.56$ | $\$4.60 \div 1\frac{1}{3} = \frac{\$4.60}{1} \div \frac{4}{3}$ <br> $= \frac{\$4.60}{1} \times \frac{3}{4}$ <br> $= \frac{\overset{\$1.15}{\cancel{\$4.60}}}{1} \times \frac{3}{\underset{1}{\cancel{4}}}$ <br> $= \frac{\$3.45}{1} = \$3.45$ |

**Compute. Round to the nearest cent when necessary.**

**1.** $\frac{1}{3}$ of \$36 ________ **2.** $\frac{1}{4}$ of \$3.20 ________ **3.** $\frac{1}{8}$ of \$8.56 ________

**4.** $\frac{1}{6}$ of \$7.20 ________ **5.** $\frac{2}{3}$ of \$52 ________ **6.** $\frac{3}{5}$ of \$72 ________

**7.** $\frac{5}{8}$ of \$12.40 ________ **8.** \$23.45 ÷ $1\frac{2}{3}$ ________ **9.** \$2.70 ÷ $1\frac{1}{5}$ ________

**10.** \$4.50 ÷ $1\frac{1}{4}$ ________ **11.** \$9.90 ÷ $1\frac{3}{8}$ ________ **12.** $\frac{3}{4}$ of \$22.50 ________

**13.** \$18.60 ÷ $2\frac{1}{4}$ ________ **14.** \$25.55 ÷ $1\frac{2}{5}$ ________ **15.** \$42 ÷ $1\frac{3}{4}$ ________

**16.** \$28.75 ÷ $\frac{5}{8}$ ________ **17.** $\frac{3}{4}$ of \$24 ________ **18.** \$24 ÷ $\frac{3}{4}$ ________

**19.** $\frac{2}{5}$ of \$50 ________ **20.** \$50 ÷ $\frac{2}{5}$ ________ **21.** \$12.32 ÷ $\frac{4}{5}$ ________

## PROBLEM SOLVING

**22.** Francis wants a camera that costs \$48. He has saved $\frac{2}{3}$ of the cost. How much has Francis saved? ______________

**23.** Sally paid \$8.55 for $4\frac{1}{2}$ yards of cloth. How much was the cloth per yard? ______________

**24.** Pedro promised to raise \$75 for a charity. He has raised $\frac{3}{5}$ of that amount. How much more money does he need to raise? ______________


Use with Lesson 7-11, text pages 246–247.
Copyright © William H. Sadlier, Inc. All rights reserved.

# Problem-Solving Strategy: Use a Diagram

Name ____________________

Date ____________________

Of 20 students, 10 are members of the Travel Club and 14 are members of the Drama Club. What part of the 20 students are members of both clubs?

Make a Venn diagram to show the facts.
Let T represent Travel Club members.
Let D represent Drama Club members.
Let B represent members of both clubs.

T B D

$T + B = 10$
$D + B = 14$
$10 + 14 = 24$
$T + B + D = 20$
$B = 24 - 20$
$B = 4$

Four students are members of both clubs.

So $\frac{4}{20}$ or $\frac{1}{5}$ of the students are in both clubs.

**PROBLEM SOLVING Do your work on a separate sheet of paper.**

1. Ten students play basketball and 16 students play baseball. If 18 students play at least one of these sports, how many students play both sports?

   ____________________

2. What part of the 18 students play both basketball and baseball?

   ____________________

3. Twelve students belong to the Science Club only, and 15 students belong to the Math Club only. If 30 students belong to at least one of these clubs, how many students belong to both clubs?

   ____________________

4. What part of the 30 students belong to both the Science and Math clubs?

   ____________________

5. There are 36 students in the History Club who are planning to take a trip. Twenty-four of these students have visited Boston and 16 have visited Philadelphia. What part of all the students have visited both cities?

   ____________________

6. A bouquet of 41 flowers is all pink and white. Nine of the flowers are both pink and white and 18 are all pink. How many flowers are all white?

   ____________________

7. There are 24 cans of vegetables in Old Mother Hubbard's cupboard. Four cans have only corn in them, 9 cans have only peas in them, and 7 cans have only carrots in them. What part of the 24 cans contain at least two vegetables?

   ____________________

8. Twenty-five members of the Coin Club collect coins from the United States or from foreign countries. Twelve members collect only foreign coins. Eight members collect both foreign and United States coins. How many members collect only coins from the United States?

   ____________________


Copyright © William H. Sadlier, Inc. All rights reserved.

# Problem Solving: Review of Strategies

Name ________________________

Date ________________________

**Solve. Do your work on a separate sheet of paper.**

**1.** Ralph's clothing company has an order for 20 skirts to be made from a bolt of fabric that has $28\frac{1}{4}$ yd of fabric. The customer wants as many flared skirts as possible, and the rest can be straight skirts. A straight skirt uses $1\frac{1}{4}$ yd of fabric, and a flared skirt uses $1\frac{1}{2}$ yd of fabric. How many of each kind of skirt can Ralph make to complete the order?

________________________

**2.** Ralph displays 9 samples of fabric on a rectangular posterboard. Each sample is a 3-inch square. He places them in 3 rows each 1 inch apart on all sides. This arrangement leaves a border of poster board $1\frac{1}{4}$ inches along the sides and $1\frac{1}{2}$ inches along the top and bottom. What are the dimensions of the posterboard?

________________________

**3.** Lynn wants to make 25 dresses. One dress uses $2\frac{2}{3}$ yd of fabric. The bolts of fabric she wants to use each have 30 yd of fabric on them. How many bolts will Lynn need to use?

________________________

**4.** A jacket requires $2\frac{1}{4}$ yd of fabric, and a skirt requires $1\frac{1}{3}$ yd of fabric. If Lynn uses the same fabric to make 15 matching jackets and skirts, how much fabric will she need?

________________________

**5.** Peter has 3 different kinds of buttons in one box. There are 124 more ivory buttons than plastic buttons and $\frac{1}{3}$ as many wooden buttons as ivory buttons. There are 72 wooden buttons. How many plastic buttons are there?

________________________

**6.** Arnie sorts letters into bins. He sorts 302 letters into the first bin, 413 letters into the second bin, and 524 letters into the third. If the pattern continues, how many letters will he put into the sixth and seventh bins?

________________________

**7.** A stick is cut so that one part is $\frac{2}{3}$ of the length of the other. How long, in inches, is the shorter part?

________________________

**8.** Ben has 204 pennies and would like to trade them for equal numbers of nickels, dimes, and quarters. What is the greatest number of each coin he can receive? How many pennies will he have left over?

________________________


Copyright © William H. Sadlier, Inc. All rights reserved.

# Graphing Sense

Name ______________________

Date ______________________

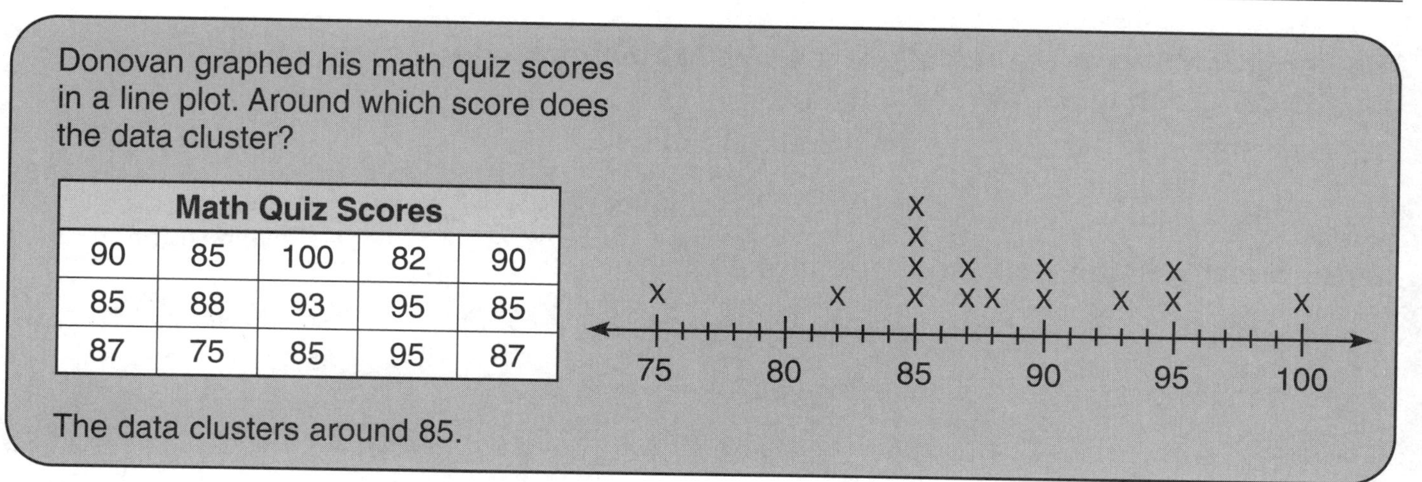

Donovan graphed his math quiz scores in a line plot. Around which score does the data cluster?

| Math Quiz Scores | | | | |
|---|---|---|---|---|
| 90 | 85 | 100 | 82 | 90 |
| 85 | 88 | 93 | 95 | 85 |
| 87 | 75 | 85 | 95 | 87 |



The data clusters around 85.

**PROBLEM SOLVING**

**Use the data in the line plot above.**

**1.** What was Donovan's most frequent score?

______________________

**2.** How many times did Donovan get a score of 75? 95?

______________________

**3.** How many times did Donovan score 90 or better?

______________________

**4.** Which scores did Donovan get only once?

______________________

**5.** Would you say that Donovan is an A (90–100), B (80–89), or C (70–79) student in math? Explain your answer.

______________________

______________________

**6.** The table at the right shows the number of CDs sold daily over a 20-day period. Use the data to make a line plot on grid paper.

| CDs Sold Daily | | | |
|---|---|---|---|
| 43 | 35 | 20 | 28 |
| 35 | 33 | 24 | 35 |
| 24 | 40 | 36 | 42 |
| 36 | 44 | 35 | 26 |
| 37 | 26 | 44 | 35 |

**7.** Make another appropriate graph to show the same data.

**8.** Write two questions that can be answered by looking at your line plot.

______________________

______________________


Copyright © William H. Sadlier, Inc. All rights reserved.

# Surveys

Name ______________________

Date ______________________

**The bar graph shows the results of a survey about making a bicycle route. Use the bar graph to answer the questions.**

1. How many people are in favor of a bicycle route?

   ______________________

2. How many more people are in favor of a bicycle route than are against it? ______________________

3. How many people in all were surveyed?

   ______________________

4. What fractional part of those surveyed are in favor of a bicycle route? ______________________

5. Would a bicycle store be an appropriate place for this survey? Explain your answer.

   ______________________

**Bicycle Route**

Response: Yes, No, Not Sure

0 10 20 30 40 50 60 70 80

**Number of People**

**The pictograph shows the results of a florist's survey about favorite flowers. Use the graph to answer the questions.**

6. How many people in all were surveyed?

   ______________________

7. How many of those surveyed did not choose either rose or tulip as their favorite flower?

   ______________________

8. Which flower was the favorite of 12 people? ______________________

9. What fractional part of those surveyed chose either iris or tulip? ______________________

10. Based on the results of the survey, about how many people out of 1000 would you expect to choose rose as their favorite flower? ______________________

11. Suppose you were the florist who took this survey. How would the results affect your orders to flower distributors?

    ______________________

    ______________________

| Favorite Flower | |
|---|---|
| Rose | ✿ ✿ ✿ ✿ |
| Iris | ✿ ✿ ✿ ½✿ |
| Daisy | ✿ ✿ |
| Carnation | ✿ ✿ ✿ |
| Tulip | ✿ ✿ ✿ ½✿ |
| Key: Each ✿ = 4 people. | |


Copyright © William H. Sadlier, Inc. All rights reserved.

# Collecting Data

Name ____________________

Date ____________________

**The frequency table shows the heights of a pediatrician's patients.**

1. Complete the table.

| Heights (meters) | Tally | Total |
|---|---|---|
| 1.21 – 1.30 | /// | |
| 1.31 – 1.40 | | 6 |
| 1.41 – 1.50 | | 7 |
| 1.51 – 1.60 | /// | |
| 1.61 – 1.70 | / | |

2. How many heights were recorded? ____________

3. Within which range are the greatest number of heights? ____________

4. How many more patients were there from 1.31 m to 1.40 m tall than were less than 1.31 m tall? ____________

5. Write a conclusion about the data in the table.

____________________________________________

**The data in the table below shows the speeds, in seconds, during the heats for the 100-meter dash at the Regional Championship Finals.**

| 100-Meter Dash: Regional Championship Finals | | | | | | | | | |
|---|---|---|---|---|---|---|---|---|---|
| 13.5 | 13.6 | 13.7 | 13.8 | 13.6 | 13.9 | 13.7 | 13.6 | 13.8 | 13.7 |
| 13.8 | 13.7 | 13.7 | 13.9 | 13.5 | 13.6 | 13.7 | 13.8 | 13.7 | 13.6 |

6. Use the data to complete the ungrouped frequency table.

| Speed (seconds) | | | | | |
|---|---|---|---|---|---|
| Tally | | | | | |
| Total | | | | | |

7. How many runners had a speed of exactly 13.8 seconds? ____________

8. Which time was run the most often? ____________

9. Which time was run by exactly 5 runners? ____________

10. How many times were recorded? ____________

11. How many more runners had a time of 13.7 seconds than a time of 13.5 seconds? ____________

12. What fractional part of the runners had times less than 13.8 seconds? Write the fraction in simplest form. ____________


Copyright © William H. Sadlier, Inc. All rights reserved.

# Range, Mean, Median, and Mode

Name ______________________

Date ______________________

Find the range, mean, median, and mode for 18, 16, 19, 16, 15, and 21.

**Range:** 21 − 15 = 6

**Mean:** 18 + 16 + 19 + 16 + 15 + 21 = 105 ⟶ 105 ÷ 6 = 17.5

**Median:** 15, 16, 16, 18, 19, 21 ⟶ 16 + 18 = 34 ⟶ 34 ÷ 2 = 17

**Mode:** 16

**Find the range, mean, median, and mode for each set of data.**

| | Data | Range | Mean | Median | Mode |
|---|---|---|---|---|---|
| **1.** | 5, 6, 7, 7, 9, 10, 12, 16 | | | | |
| **2.** | 25, 30, 40, 25, 50, 40 | | | | |
| **3.** | 72, 80, 120, 98, 78, 84, 77 | | | | |
| **4.** | 8.5, 8.5, 10.1, 4.8, 4.6, 8.5 | | | | |
| **5.** | 7.2, 6.1, 8.3, 7.2, 7.5, 7.6, 5.1 | | | | |
| **6.** | $8.20, $1.15, $10.75, $5.50 | | | | |
| **7.** | $2.45, $2.15, $2.45, $2.25, $2.40 | | | | |

**PROBLEM SOLVING Use the box-and-whisker plot for exercises 9-12.**

**8.** Andre's test scores were 85, 84, 53, 86, 80, 82, 87, and 53. Would the mean, median, or mode best describe these scores? Explain.

______________________________________________

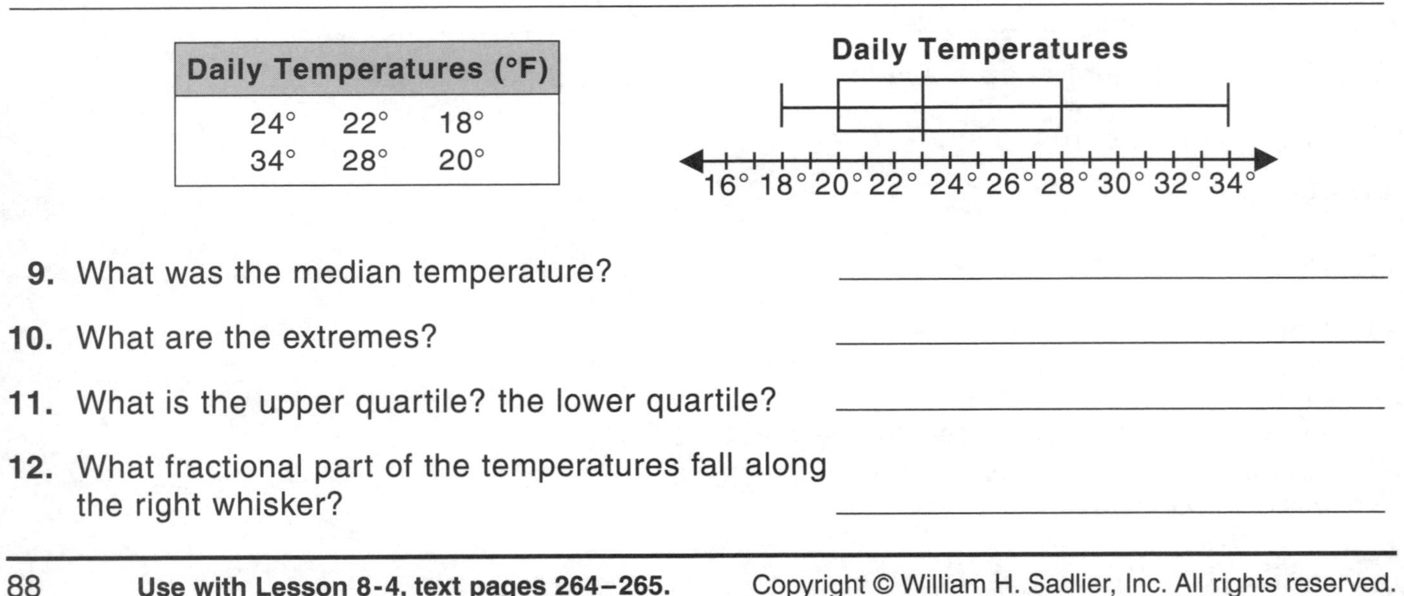

| Daily Temperatures (°F) | | |
|---|---|---|
| 24° | 22° | 18° |
| 34° | 28° | 20° |



**9.** What was the median temperature? ______________

**10.** What are the extremes? ______________

**11.** What is the upper quartile? the lower quartile? ______________

**12.** What fractional part of the temperatures fall along the right whisker? ______________

 **Use with Lesson 8-4, text pages 264–265.** Copyright © William H. Sadlier, Inc. All rights reserved.

# Stem-and-Leaf Plot

Name ______________________

Date ______________________

**The stem-and-leaf plot shows daily high temperatures during July. Use the plot to answer questions 1-4.**

| Stem | Leaf |
|---|---|
| 6 | 9 |
| 7 | 3 5 |
| 8 | 2 4 8 8 9 |
| 9 | 4 5 5 |

1. The temperatures for how many days are shown?
______________________

2. What was the highest temperature?
______________________

3. For how many days was the temperature over 80°?
______________________

4. What is the range, median, and mode of the data?
______________________

**The data shows daily low temperatures (°F) for Miami, Florida during the month of February. Complete the stem-and-leaf plot. Then answer questions 6-8.**

5.

| Stem | Leaf |
|---|---|
| 4 | |
| 5 | |
| 6 | |

| | | | | | | | |
|---|---|---|---|---|---|---|---|
| 48° | 53° | 59° | 56° | 58° | 59° | 49° | 60° |
| 60° | 46° | 62° | 55° | 59° | 57° | 54° | 61° |

6. What is the range, median, and mode of the data? ______________________
7. How many days was the temperature below 50°F? ______________________
8. Write a statement that summarizes what the stem-and-leaf plot shows. ______________________

**The data shows the heights (numbers of floors) of some tall buildings in Detroit, Michigan. Use the data to answer questions 9-12.**

| | | | | | | | | | | |
|---|---|---|---|---|---|---|---|---|---|---|
| 47 | 40 | 71 | 28 | 28 | 40 | 39 | 35 | 32 | 32 | 40 |
| 38 | 32 | 27 | 27 | 27 | 28 | 25 | 34 | 26 | 25 | 19 |

9. Make a stem-and-leaf plot of the data.

10. What is the range, median and mode of the data?
______________________

11. How many buildings have a height of 40 stories or more? ______________________

12. Write a statement that summarizes what the stem-and-leaf plot shows about the tallest building.
______________________


Copyright © William H. Sadlier, Inc. All rights reserved.

# Working with Graphs and Statistics

Name ____________________

Date ____________________

**Use the line graph for questions 1-2.**

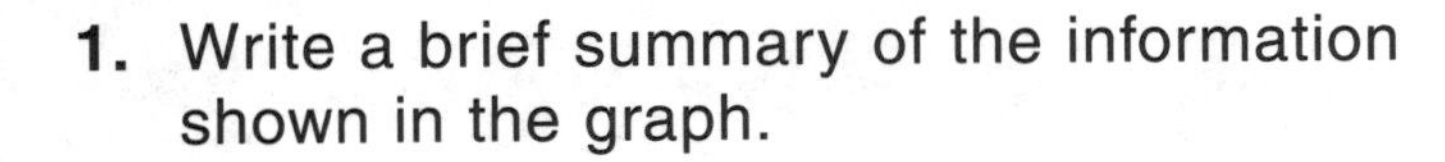

1. Write a brief summary of the information shown in the graph.

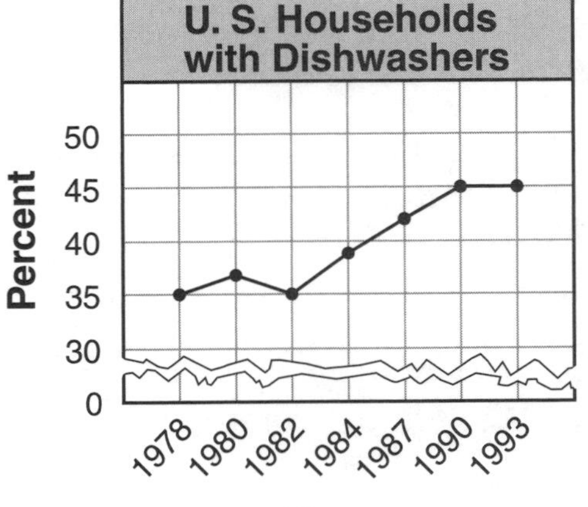


2. What trends, if any, do you see in the data?

**Use the information in the table for exercises 3-5.**

| Final Exam Math Scores in Mr. Wong's Class | | | | | | | |
|---|---|---|---|---|---|---|---|
| 68 | 76 | 79 | 99 | 78 | 100 | 89 | 85 |
| 95 | 82 | 82 | 85 | 83 | 79 | 75 | 88 |
| 98 | 85 | 56 | 93 | 65 | 86 | 98 | 79 |

3. Complete the frequency table.

| Math Scores | Tally | Total |
|---|---|---|
| below 60 | | 1 |
| 60 – 69 | | 2 |
| 70 – 79 | | 6 |
| 80 – 89 | | 9 |
| 90 – 100 | | 6 |

4. What is the range, median, and mode of this data?

Range: ____________________

Median: ____________________

Mode: ____________________

5. Which statistic—range, median, or mode most accurately describes the typical math score in Mr. Wong's class? Explain.


Copyright © William H. Sadlier, Inc. All rights reserved.

# Making Line Graphs

Name ______________________

Date ______________________

**Complete the graph to show the data in the table.**

1.

| Profits Earned by the Beach Company | |
|---|---|
| **Month** | **Profit** |
| January | $30,000 |
| March | $25,000 |
| May | $50,000 |
| July | $65,000 |
| September | $35,000 |
| November | $35,000 |

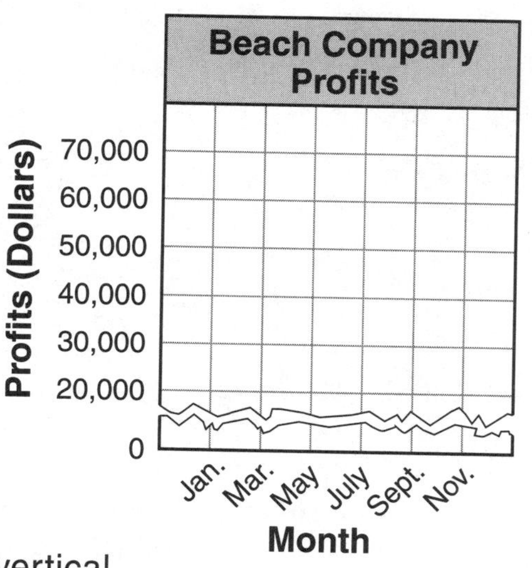


2. How much money does each unit on the vertical scale represent? ______________________

3. What trend does the graph show? ______________________

4. By how much did profits decrease from July to November? ______________________

5. What is the range of the profits? the mean profit? ______________________

6. Between which two months was there the greatest increase in profits? the greatest decrease? ______________________

**Make a line graph for the set of data below.**

7.

| Snowfall in Combsville | |
|---|---|
| **Year** | **Amount (in inches)** |
| 1966 | 20 |
| 1967 | 23 |
| 1968 | 19.5 |
| 1969 | 21.5 |
| 1970 | 17 |
| 1971 | 15.5 |

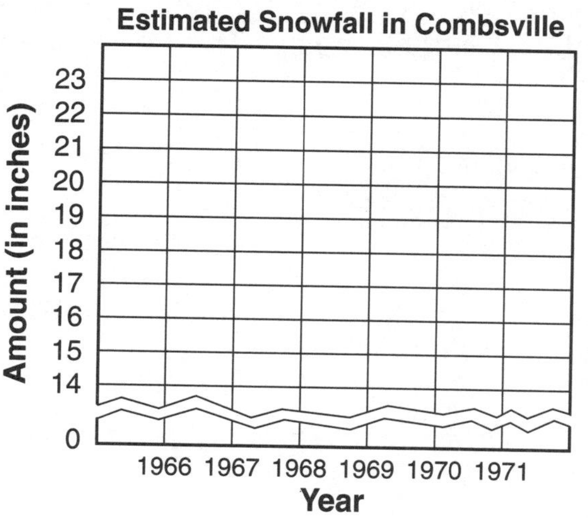


8. What scale did you use for your graph? Why?

______________________________________________

______________________________________________


Copyright © William H. Sadlier, Inc. All rights reserved.

# Analyzing Line Graphs

Name ______________________

Date ______________________

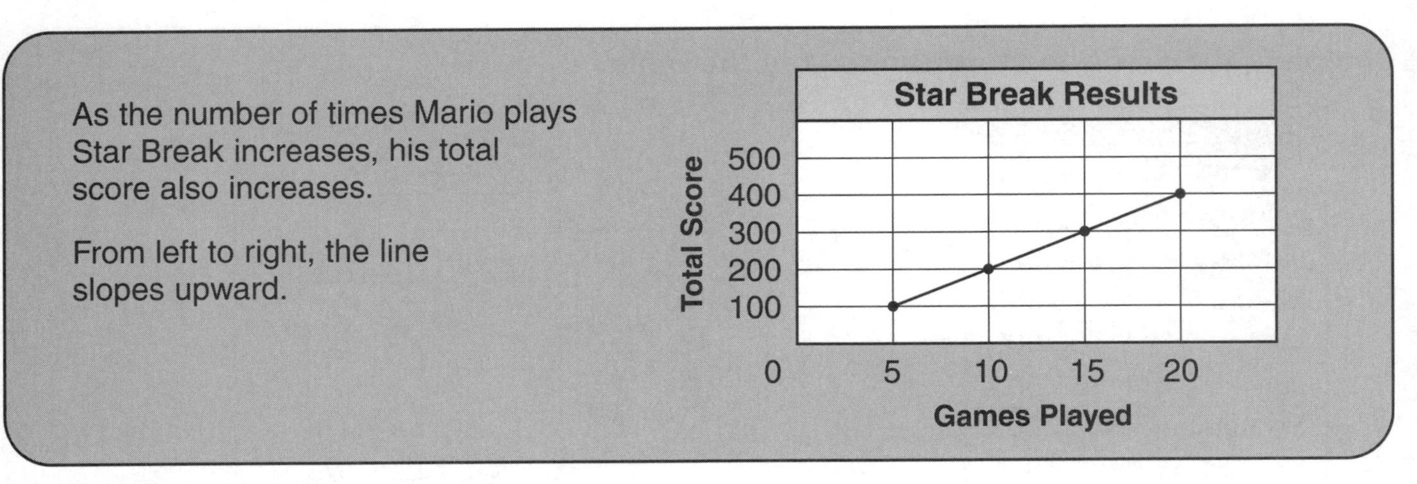


As the number of times Mario plays Star Break increases, his total score also increases.

From left to right, the line slopes upward.

**PROBLEM SOLVING**

**1.** The first 75 games Mario played, his time for a complete game decreased by 10 seconds for every 5 games. Draw a line graph of the data.

| Games Played | 5 | 10 | 15 | 20 | 25 |
|---|---|---|---|---|---|
| Number of Seconds | 180 | 170 | 160 | 150 | 140 |

**2.** For exercise 1, does the line slope upward or downward? Explain why.

______________________________________________

______________________________________________

**3.** Shamika's watch gains 5 minutes every 18 hours. Complete the table below. Then draw a line graph of the data.

| Number of Hours | 18 | 36 | | | 90 |
|---|---|---|---|---|---|
| Minutes Gained | 5 | | 15 | | |

**4.** For exercise 3, does the line slope upward or downward? Explain why.

______________________________________________

______________________________________________

**5.** How many minutes will Shamika's watch gain in 162 hours?

______________________________________________

 **Use with Lesson 8-8, text pages 272–273.** Copyright © William H. Sadlier, Inc. All rights reserved.

# Double Line and Double Bar Graphs

Name ____________________

Date ____________________

## PROBLEM SOLVING

**Use the double line graph for problems 1-4.**

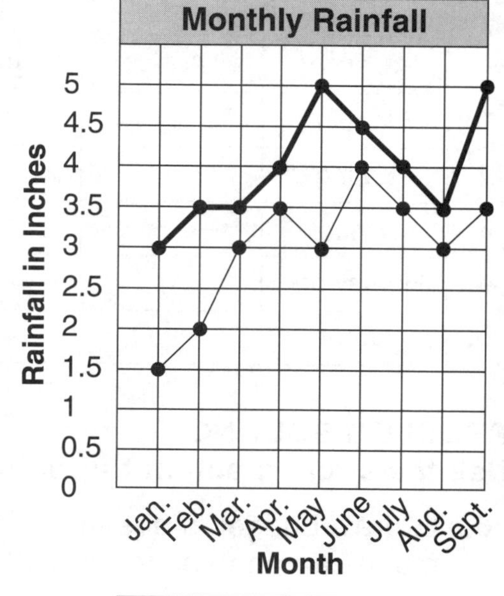

1. How much more rain fell in Houston than in St. Louis during September? ____________

2. What can you say about the rainfall in St. Louis between January and April? ____________

3. In which city did the amount of rainfall stay the same for two consecutive months? ____________

4. When was the difference between the amount of rainfall in Houston and in St. Louis the greatest?

   ____________

**Use the double bar graph for problems 5-8.**

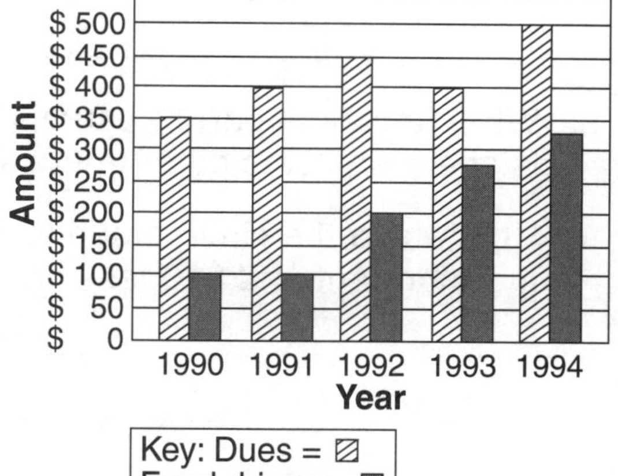


Key: Dues = ▨
Fund drives = ■

5. How much more money was made in dues than in fund drives during 1991? ____________

6. During which year was the most money made in dues? in fund drives? ____________

7. During which year did the club earn the greatest total amount? the least total amount?

   ____________

8. During which year was there the least difference between earnings from dues and from fund drives? the most difference? ____________

**Use grid paper for problems 9-10.**

9. Make a double line graph for this table of book sales at Betty's Book Store.

| Month | Children's | Adult's |
|---|---|---|
| Apr. | 130 | 150 |
| May | 125 | 140 |
| June | 120 | 145 |
| July | 95 | 120 |
| Aug. | 100 | 110 |

10. Make a double bar graph for this table of high and low math scores.

| Test | High Score | Low Score |
|---|---|---|
| Test 1 | 95 | 60 |
| Test 2 | 89 | 65 |
| Test 3 | 90 | 70 |
| Test 4 | 95 | 75 |
| Test 5 | 85 | 68 |


Copyright © William H. Sadlier, Inc. All rights reserved.

# Interpreting Circle Graphs

Name ______________________

Date ______________________

**Use the circle graph at the right to complete this table.**

| | Activity | Fraction | Hours |
|---|---|---|---|
| **1.** | Sleeping | | |
| **2.** | Eating | | |
| **3.** | School | | |
| **4.** | Homework | | |
| **5.** | Other | | |

**Ron's Day**

Eating $\frac{1}{12}$

Homework $\frac{1}{8}$

Sleeping $\frac{3}{8}$

School $\frac{1}{4}$

Other $\frac{1}{6}$

**PROBLEM SOLVING**
**Use the circle graph at the right for problems 6-9.**

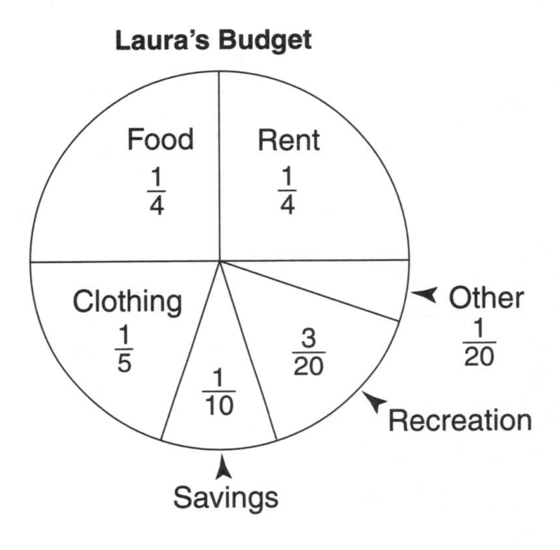


**6.** Laura's budget is based on $200 per week. How much can she spend on clothing each week?

______________________

**7.** For which two items has Laura budgeted the same amount of money? how much money?

______________________

**8.** What fractional part of Laura's budget is spent on items other than savings? ____________

**9.** How would the circle graph look if Laura spent $10 more on rent and $10 less on food?

______________________

______________________

**Use the circle graph at the right for problems 10-13.**

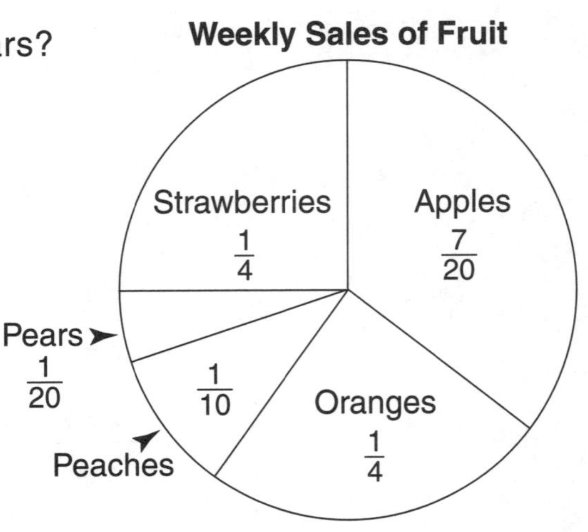


**10.** What fractional part of the sales is peaches and pears?

______________________

**11.** What fractional part of the fruit sold is not apples?

______________________

**12.** If 600 bushels of fruit are sold weekly, how many bushels are oranges? ____________

**13.** If 600 bushels of fruit are sold weekly, how many bushels are pears? ____________

 **Use with Lesson 8-10, text pages 276–277.** Copyright © William H. Sadlier, Inc. All rights reserved.

# Probability

Name ______________________

Date ______________________

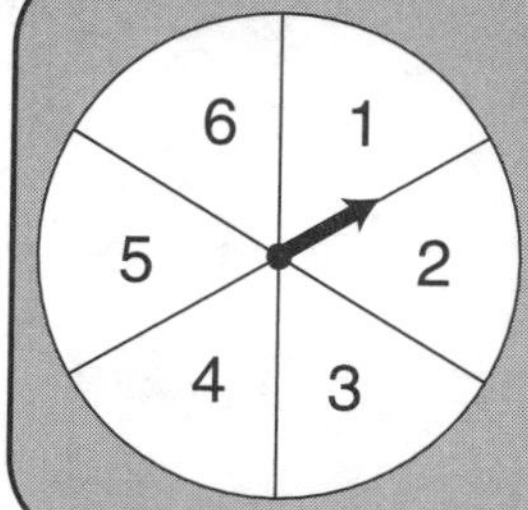


$P(3) = \frac{\text{number of favorable outcomes}}{\text{number of possible outcomes}} = \frac{1}{6}$ ← one favorable outcome: 3 ← six outcomes in all

$P(3 \text{ or } 5) = P(3) + P(5) = \frac{1}{6} + \frac{1}{6} = \frac{2}{6} = \frac{1}{3}$

$P(\text{not } 3) = \frac{5}{6}$ five favorable outcomes: 1, 2, 4, 5 and 6

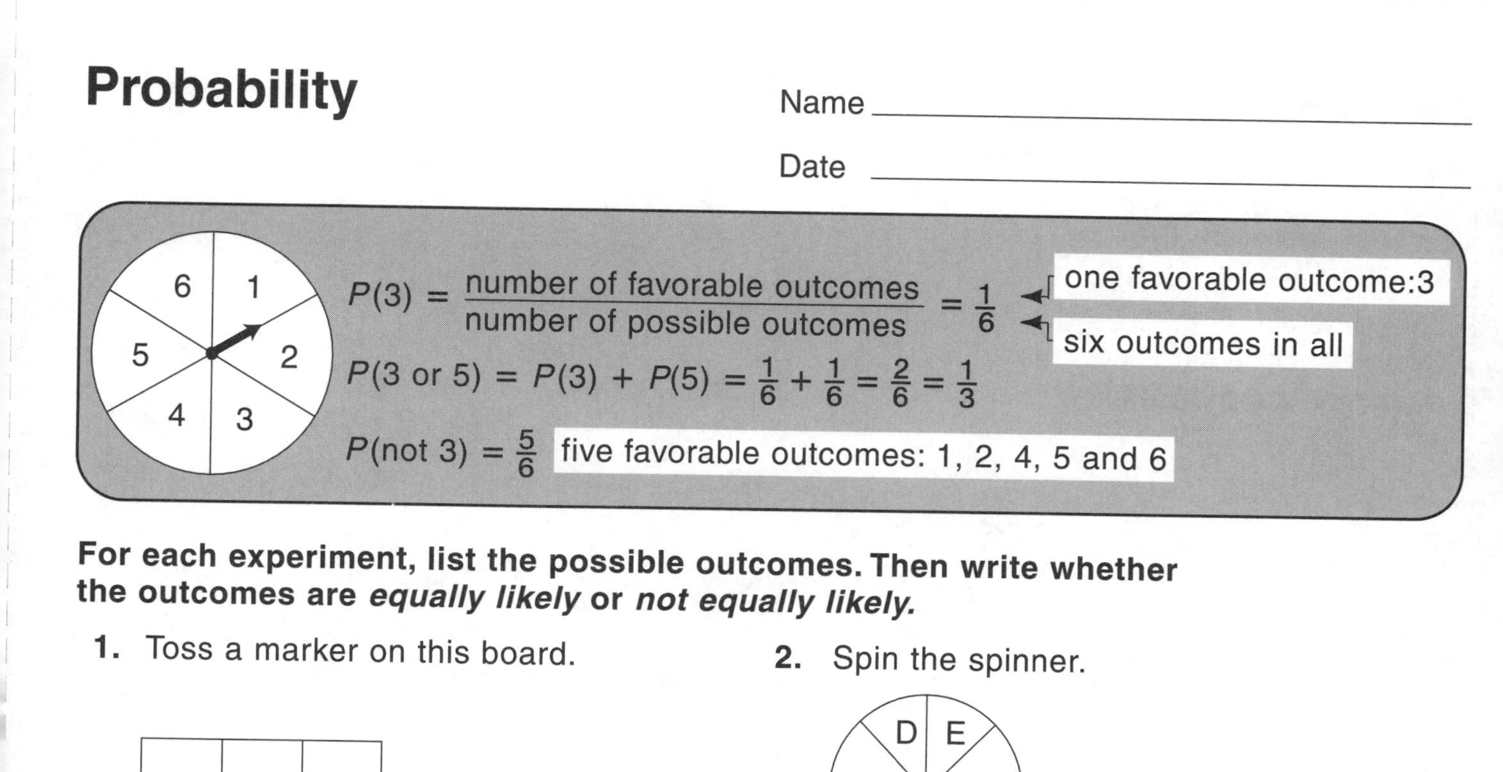

**For each experiment, list the possible outcomes. Then write whether the outcomes are *equally likely* or *not equally likely.***

1. Toss a marker on this board.

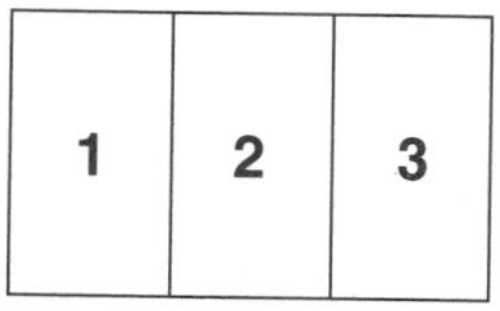


______________________

2. Spin the spinner.

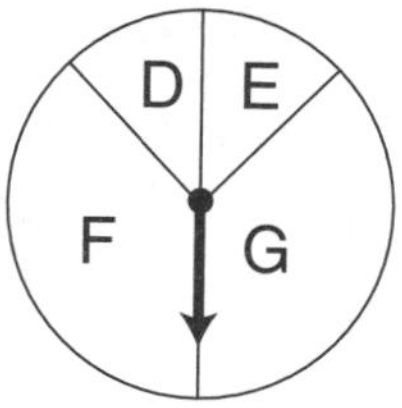


______________________

**Use the spinner to find the probability of each event.**

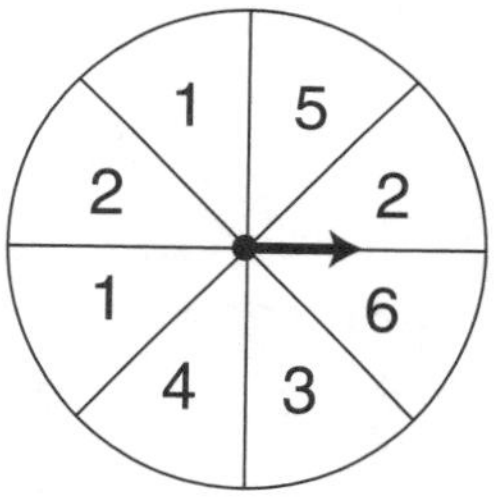


3. $P(2)$ ________
4. $P(3)$ ________
5. $P(4)$ ________
6. $P(1 \text{ or } 2)$ ________
7. $P(\text{odd or even number})$ ________
8. $P(\text{not } 2)$ ________
9. $P(\text{not } 5)$ ________
10. $P(\text{even number})$ ________
11. $P(\text{odd number})$ ________
12. $P(1 \text{ or } 6)$ ________

**One crayon is picked at random from a box containing 2 red crayons, 3 green crayons, 1 purple crayon, 1 blue crayon, and 1 yellow crayon. Find each probability.**

13. $P(\text{yellow})$ ________
14. $P(\text{green})$ ________
15. $P(\text{pink})$ ________
16. $P(\text{not blue})$ ________
17. $P(\text{red or green})$ ________
18. $P(\text{not red})$ ________
19. $P(\text{blue or green})$ ________
20. $P(\text{red, green, purple, blue, or yellow})$ ________

**Use these six cards to find the probability of each event.**

| A | B | D | E | D | C |
|---|---|---|---|---|---|

21. $P(A)$ ________
22. $P(B)$ ________
23. $P(D)$ ________
24. $P(C \text{ or } D)$ ________
25. $P(C \text{ or } E)$ ________
26. $P(F)$ ________
27. $P(\text{not } A)$ ________
28. $P(\text{not } D)$ ________

 Copyright © William H. Sadlier, Inc. All rights reserved.

# Compound Events

Name ______________________________

A B C

**Independent Events**

Experiment 1: Draw a card. Replace it. Draw a second card. How many outcomes are there? What is *P*(A, B)?

**Dependent Events**

Experiment 2: Draw a card. Do not replace it. Draw a second card. How many outcomes are there? What is *P*(A, B)?

To find the probability that one event and then another will occur:

- Find the probability of each event.
- Multiply the two probabilities.

| Experiment 1 | Experiment 2 |
|---|---|
| $P(A,B) = \underline{?}$ | $P(A,B) = \underline{?}$ |
| $P(A) = \frac{1}{3}$, $P(B) = \frac{1}{3}$ | $P(A) = \frac{1}{3}$, $P(B) = \frac{1}{2}$ |
| $P(A,B) = \frac{1}{3} \times \frac{1}{3} = \frac{1}{9}$ | $P(A,B) = \frac{1}{3} \times \frac{1}{2} = \frac{1}{6}$ |

**Draw a tree diagram for each on a separate sheet of paper.**
**Write the number of all possible outcomes.**

**1.** Roll a 1-6 number cube and spin the spinner.

4 2 3 A B

______________________

**2.** Spin the spinner and select a card.

A B C 1 2 3 4

______________________

**3.** You are choosing a drink and a sandwich from a menu offering 3 kinds of drinks and 5 kinds of sandwiches.

______________________

**4.** You are choosing an outfit from a selection of 3 shirts, 2 pairs of pants, and 3 sweaters.

______________________

**Find each probability. Use the experiment in exercise 1.**

**5.** *P*(3, B) ______ **6.** *P*(even, A) ______ **7.** *P*(2, A or B) ______ **8.** *P*(not 1, B) ______

**Find each probability. Use the experiment in exercise 2.**

**9.** *P*(A, 3) ______ **10.** *P*(B, even) ______ **11.** *P*(C, 1, 2, or 3) ______ **12.** *P*(not A, odd) ______

**Find the probability: (a) If the first choice is replaced; and (b) if the first choice is not replaced.**

Choose a marble from a bag having 3 green, 2 red, and 4 blue marbles. Then choose a second marble.

**13.** *P*(G, R) **14.** *P*(B,G) **15.** *P*(R,B)

______ ______ ______

 **Use with Lesson 8-12, text pages 280–281.** Copyright © William H. Sadlier, Inc. All rights reserved.

# Predictions

Name ____________________

Date ____________________

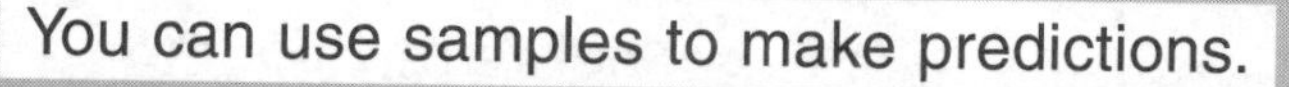


Out of 25 students surveyed, 15 said they would vote for Liz for class president. How many students would you expect to vote for Liz out of 1000 students?

$$\frac{15}{25} = \frac{3}{5} \qquad \frac{3}{\cancel{5}_1} \times \frac{\cancel{1000}^{200}}{1} = 600$$

Based on the survey, you can expect about 600 students to vote for Liz.

**You can use probability to make predictions.**

In 500 spins, predict how many times the spinner will land on 2.

$$\frac{1}{\cancel{5}_1} \times \frac{\cancel{500}^{100}}{1} = 100$$

Based on a probability of $\frac{1}{5}$, you can predict the spinner will land on 2 about 100 times.

**In 1000 spins, predict the number of times the spinner above would land on each of the following.**

**1.** number $< 4$ ____ **2.** number $> 4$ ____ **3.** odd number ____ **4.** even number ____

**PROBLEM SOLVING**

**5.** Out of 25 light bulbs tested, 1 was found to be defective. Out of 2000 light bulbs, how many would you expect to be defective? ____________

**6.** In a sample of 200 people, 150 said they would vote for Davis. How many people out of 6000 would you expect to vote for Davis in the election? ____________

**7.** About how many fours would you expect to get if you rolled a 1-6 number cube 100 times? ____________

**8.** Tom reached into a bag containing 500 marbles. He pulled out 16 marbles, and 4 of them were blue. How many marbles of the 500 would you expect to be blue? ____________

**9.** In 2000 spins, predict the number of times the spinner at the right will land on B. ____________

**10.** In 2000 spins, predict the number of times the spinner at the right will land on D. ____________

**The bar graph shows the fruit drink preferences of 100 people.**

**11.** In a population of 1000 people, how many would you expect to choose each type of drink?

____________________

**12.** Suppose apple was not given as a choice, and 5 more people chose grape and 10 more people chose orange. How many people out of 1000 would you expect to choose each type of drink?

____________________

 Copyright © William H. Sadlier, Inc. All rights reserved. 

# Misleading Graphs and Statistics

Name____________________

Date____________________

**PROBLEM SOLVING**

**The table and graph below show the heights of buildings.**

| Building | Height (floors) |
|---|---|
| InterFirst Plaza | 55 |
| World Trade Center | 110 |
| Texas Commerce Tower | 75 |
| John Hancock | 100 |

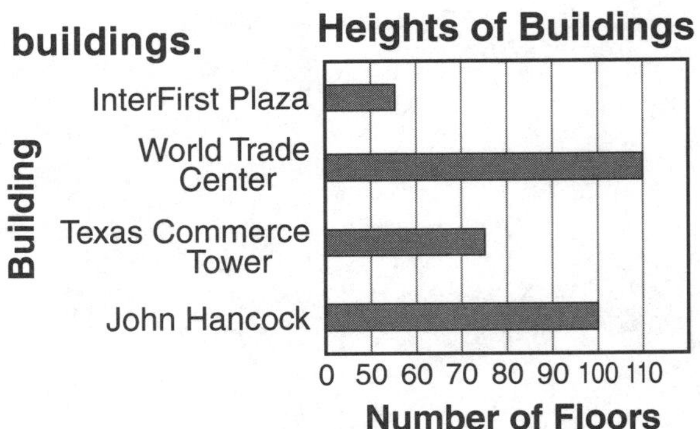


1. According to the graph which building appears to be about half as tall as the World Trade Center? ____________________

2. Which building actually is about half as tall? ____________________

3. What makes this graph misleading? ____________________

**The graphs show the number of books sold over a 5-month period.**

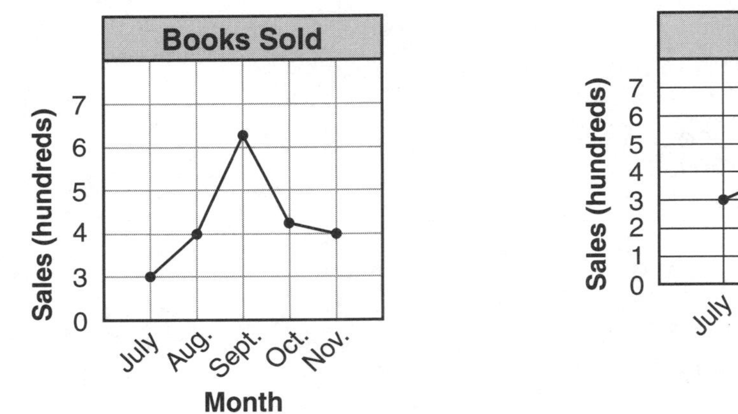


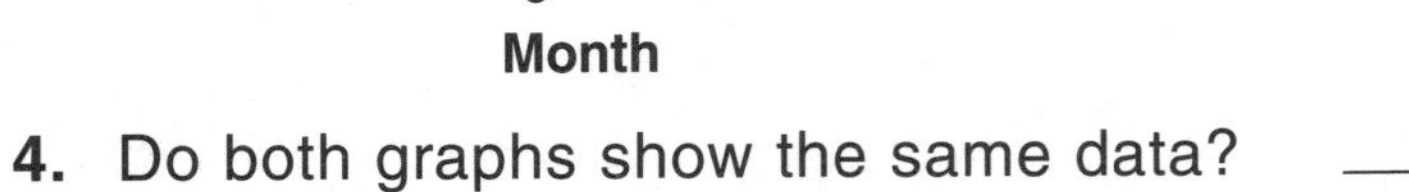

4. Do both graphs show the same data? ____________________

5. How are the graphs different? ____________________

**The graph at the right shows the attendance at the Hiker's Club yearly business meeting.**

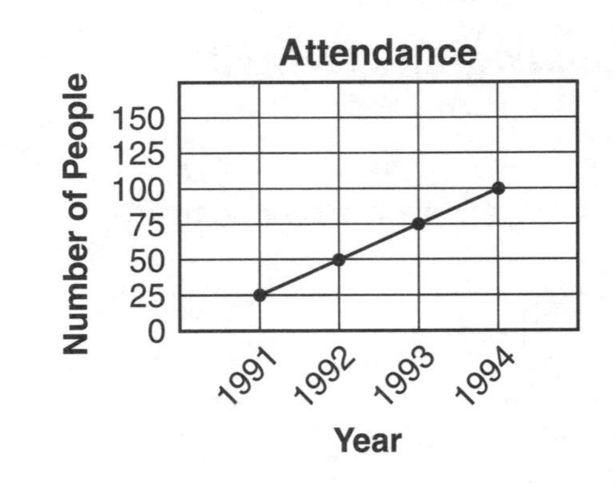


6. Was there a steady increase or a steady decrease in attendance? ____________________

7. Does the graph give the impression that the change in attendance over the last 4 years is small or great? Which is it? ____________________

8. How would you change the graph to make it represent the data better? ____________________


Copyright © William H. Sadlier, Inc. All rights reserved.

# Problem-Solving Strategy: Organized List

Name ______________________

Date ______________________

A bag holds a red marble, a white marble, and a blue marble.
Without looking, you reach in, pick a marble, note its color, and put it back.
You do this again. How many different color combinations are possible?
Make an organized list.

| **First Pick** | *red* | *red* | *red* | *white* | *white* | *white* | *blue* | *blue* | *blue* |
|---|---|---|---|---|---|---|---|---|---|
| **Second Pick** | *red* | *white* | *blue* | *red* | *white* | *blue* | *red* | *white* | *blue* |

There are 9 possible color combinations.

**PROBLEM SOLVING Do your work on a separate sheet of paper.**

1. Zev has 5 T-shirts in a drawer: 1 blue, 1 white, 1 red, 1 green, and 1 yellow. Without looking, he reaches in and takes out one shirt, and then he reaches in and takes out another shirt. How many different ways can he take out two shirts?

   ______________________

2. Leda needs to choose one letter and one digit from the following: X, Y, Z, 0, 1, 2, 3. She must place the letter first to form a 2-character code. How many different 2-character codes are there from which Leda can choose?

   ______________________

3. How many different 3-digit numbers can you make with the digits 1, 2, and 3 if repetition is permitted?

   ______________________

4. Ben has 5 tulip bulbs and 3 window boxes. In how many ways can he plant the bulbs so that each window box has at least one bulb?

   ______________________

5. Al has a red shirt, a white shirt, and a yellow shirt. He has a pair of brown slacks, a pair of black slacks, and a pair of blue slacks. How many different outfits can he make?

   ______________________

6. In a multiplication game you spin the spinner twice and find the product of the two numbers. How many different products are possible if the spinner is divided into 4 equal sections numbered 1, 2, 3, 4 respectively?

   ______________________

7. Victoria is spending her vacation in Washington, D. C. She wants to visit the Capitol Building, the White House, the Lincoln Memorial, and the Washington Monument. In how many different orders can she visit all four places?

   ______________________

8. A library has a front door, a door on the left side of the building, another door on the right side, and two rear doors. Victor can enter the library through the front door and leave through any door. In how many different ways can he do this?

   ______________________

Use with Lesson 8-16, text pages 288–289.
Copyright © William H. Sadlier, Inc. All rights reserved.

# Congruent Segments and Angles

Name ______________________

Date ______________________

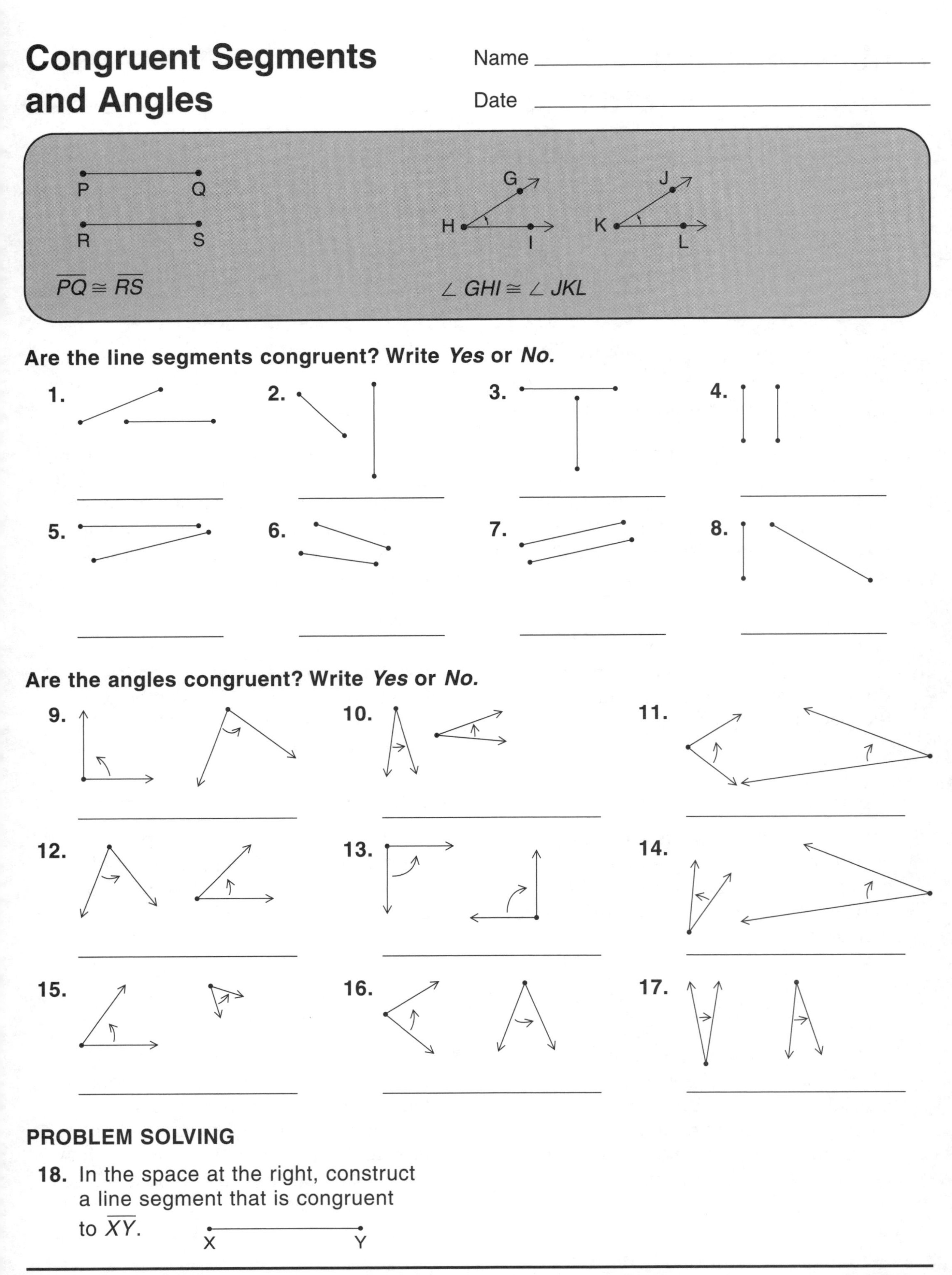


**Are the line segments congruent? Write *Yes* or *No*.**

**1.** ______ **2.** ______ **3.** ______ **4.** ______

**5.** ______ **6.** ______ **7.** ______ **8.** ______

**Are the angles congruent? Write *Yes* or *No*.**

**9.** ______ **10.** ______ **11.** ______

**12.** ______ **13.** ______ **14.** ______

**15.** ______ **16.** ______ **17.** ______

**PROBLEM SOLVING**

**18.** In the space at the right, construct a line segment that is congruent to $\overline{XY}$.

X Y

 **Use with Lesson 9-1, text pages 296–297.** Copyright © William H. Sadlier, Inc. All rights reserved.

# Constructing Perpendicular Lines

Name ______________________________

Date ______________________________

A D C B E

$\overleftrightarrow{AB}$ and $\overleftrightarrow{CE}$ are perpendicular lines.

$\overleftrightarrow{AB} \perp \overleftrightarrow{CE}$

**Use a straightedge and compass. Construct a line that is perpendicular to the given line at the given point.**

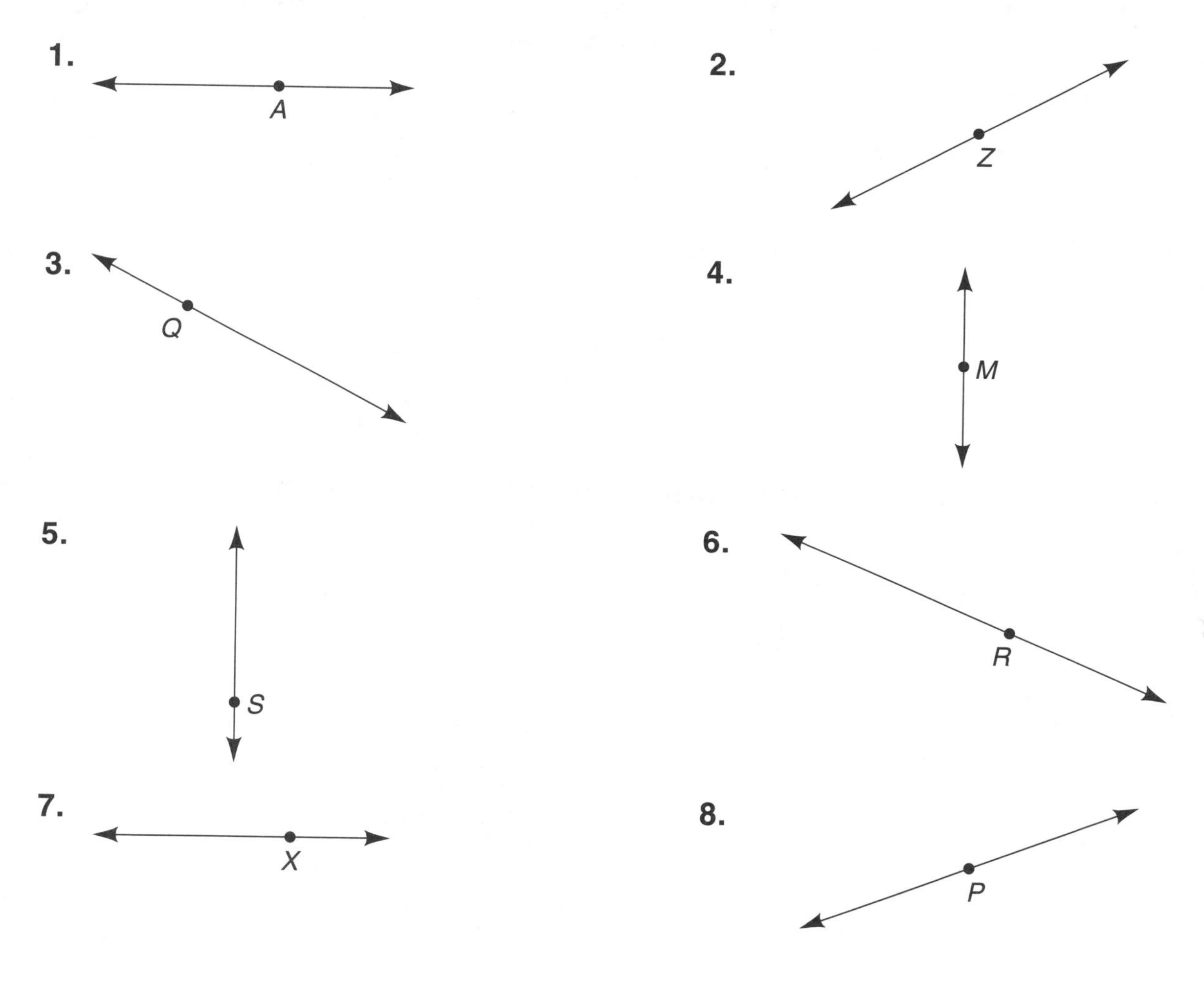


## PROBLEM SOLVING

9. How many angles are formed by two perpendicular lines? ______________________________

10. Are the angles formed by two perpendicular lines congruent? What kind of angles are they? ______________________________

 Copyright © William H. Sadlier, Inc. All rights reserved. 

# Measuring and Drawing Angles

Name ____________________

Date ____________________

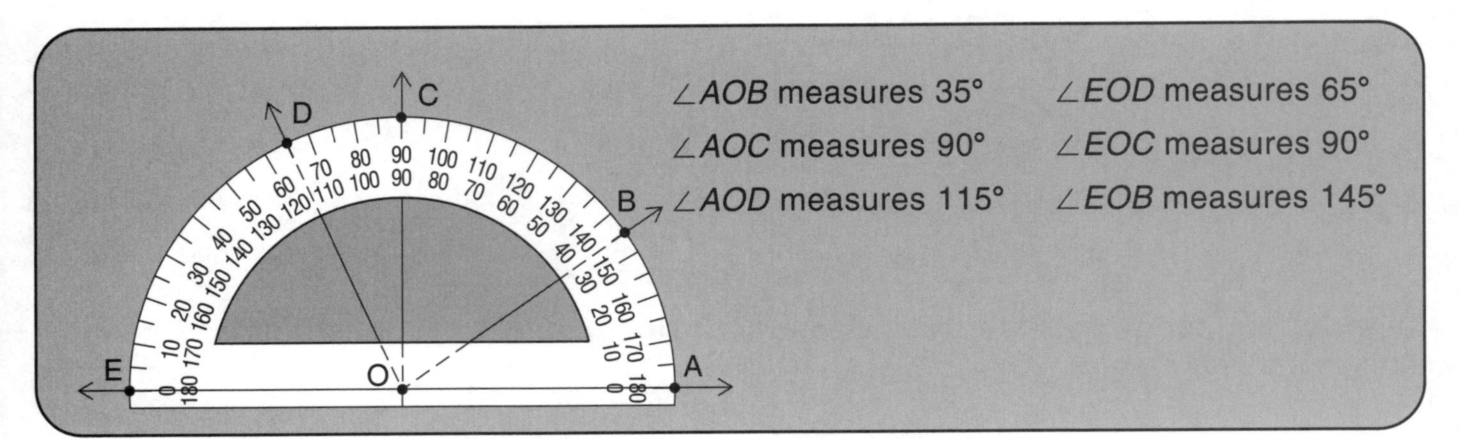


**Give the measure of each angle.**

**1.** ∠AOB ______ **2.** ∠AOC ______

**3.** ∠AOD ______ **4.** ∠FOE ______

**5.** ∠FOD ______ **6.** ∠FOB ______

D C E B F A O

**Estimate the measure of each angle. Then use a protractor to find the exact measure.**

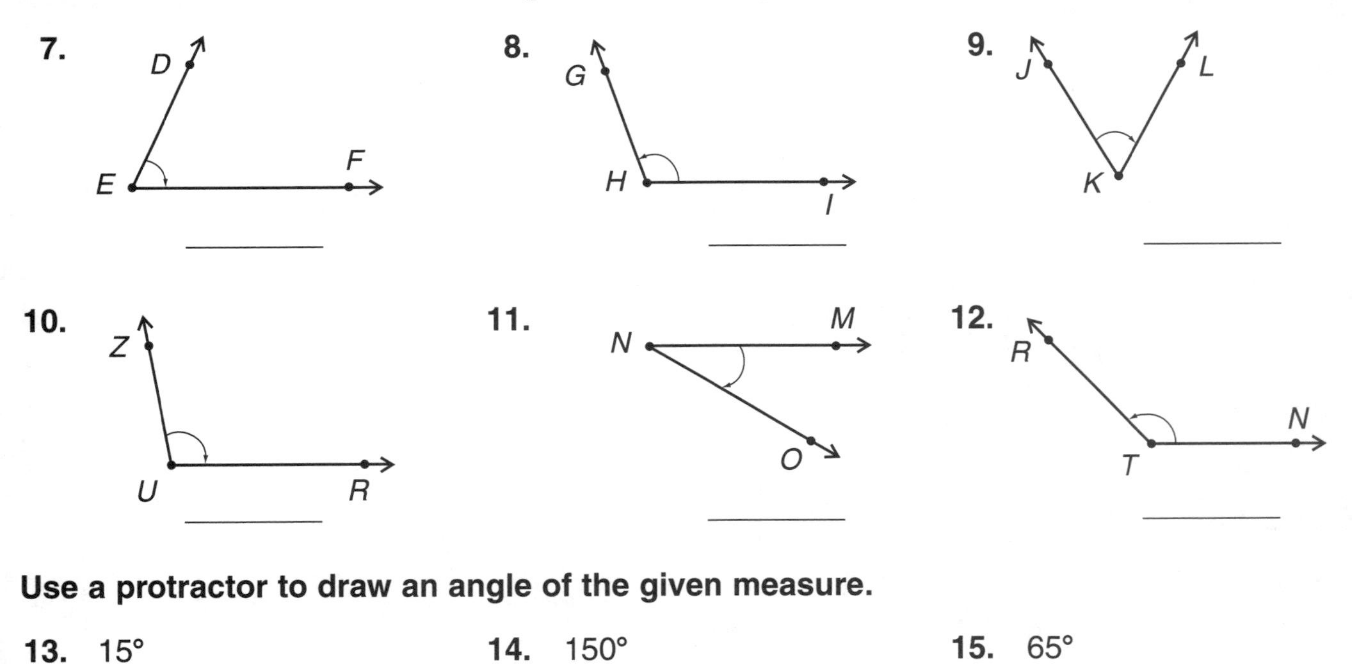


**Use a protractor to draw an angle of the given measure.**

**13.** 15° **14.** 150° **15.** 65°

 Copyright © William H. Sadlier, Inc. All rights reserved.

# Classifying Angles

Name ______________________

Date ______________________

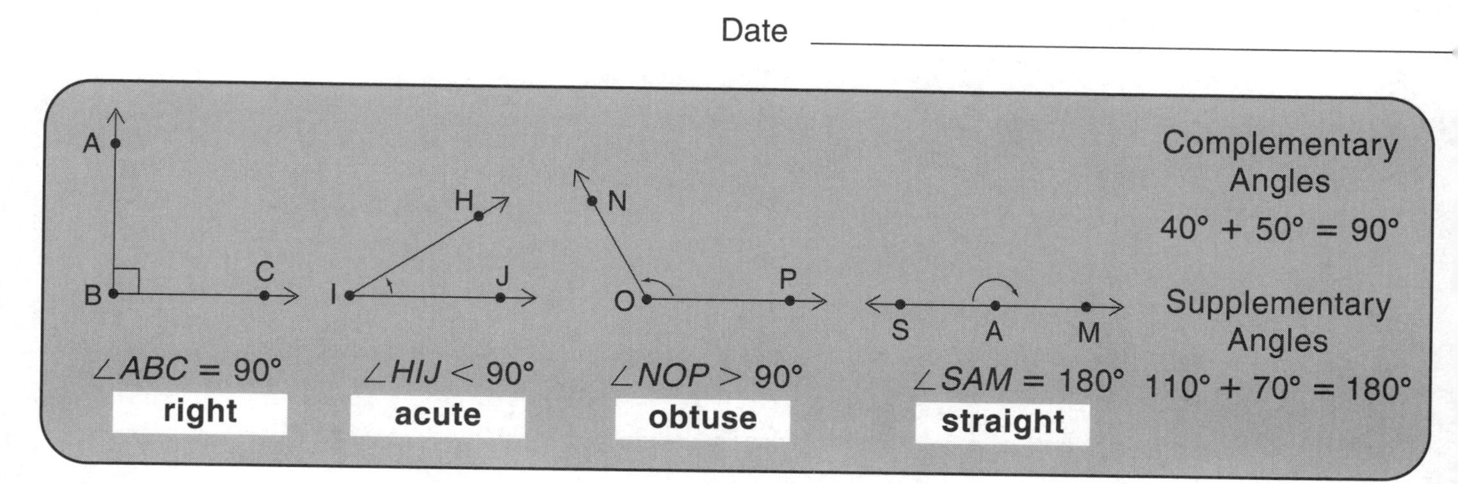


**Classify each angle as *right, acute, obtuse,* or *straight.* Use a protractor to check.**

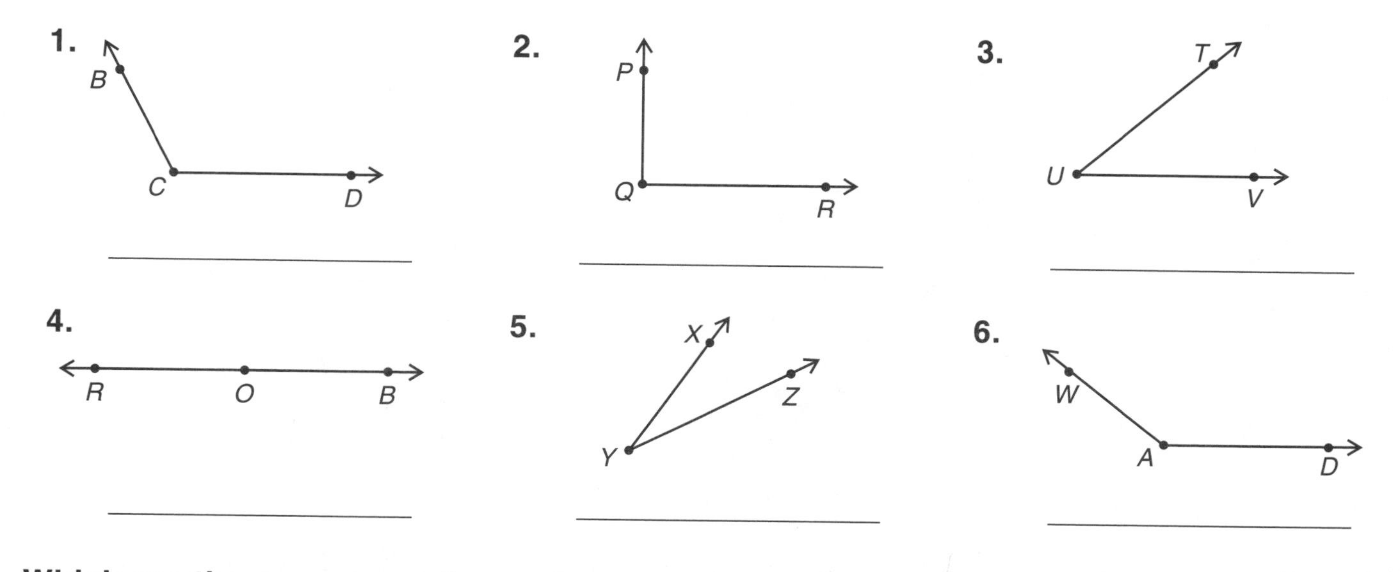


**Which are the measures of complementary angles? supplementary angles? neither?**

7. 45°, 45° ______________
8. 73°, 107° ______________
9. 35°, 65° ______________
10. 170°, 10° ______________
11. 87°, 83° ______________
12. 62°, 28° ______________
13. 89°, 1° ______________
14. 20°, 70° ______________

**Draw the angles described.**

15. Two complementary angles that are congruent and share a common side
16. Two supplementary angles that share a common side

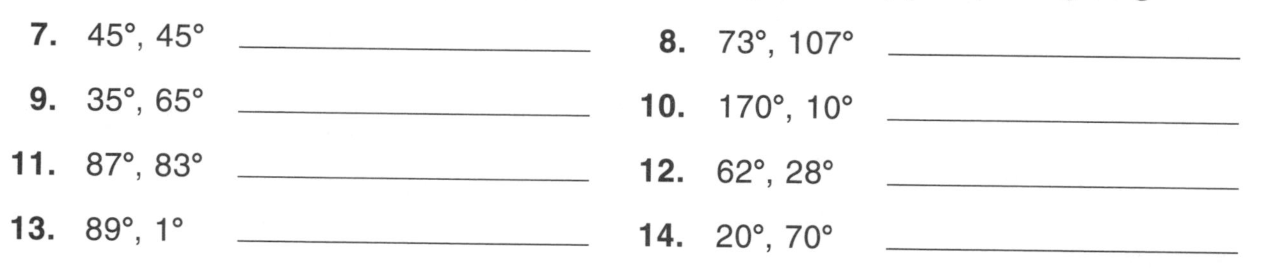


Copyright © William H. Sadlier, Inc. All rights reserved.

# Constructions With Angles

Name ______________________________

Date ______________________________

Construct $\angle DEF$ congruent to $\angle ABC$.

A, P, B, Q, C, E, F, E, D, F

$\angle DEF \cong \angle ABC$

Bisect $\angle XYZ$.

X, Y, Z, X, A, Y, B, Z, X, A, P, Y, B, Z

$\overrightarrow{YP}$ is the bisector of $\angle XYZ$.

**Construct an angle congruent to each angle below.**

**1.** **2.** **3.**

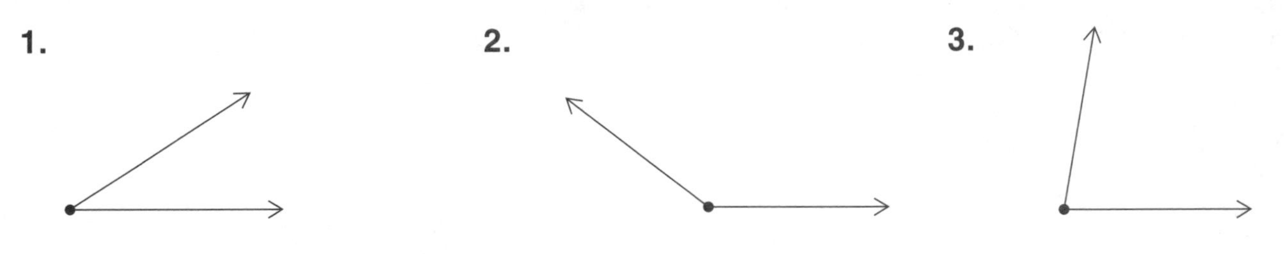

**Name the bisector of each angle and the two congruent angles formed.**

**4.** **5.** **6.**

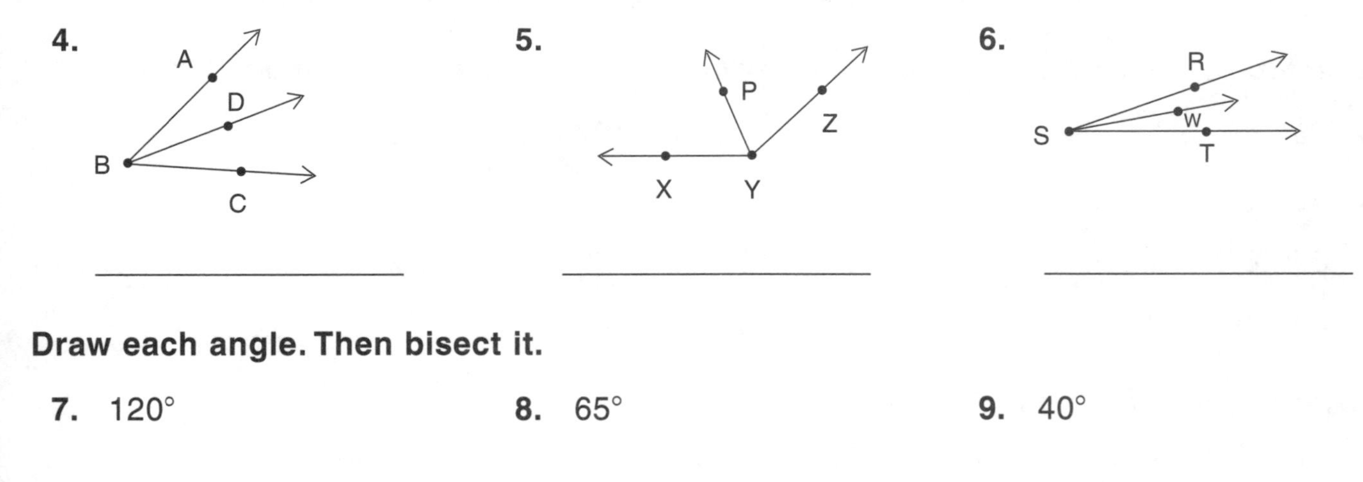


______________________ ______________________ ______________________

**Draw each angle. Then bisect it.**

**7.** 120° **8.** 65° **9.** 40°

 **Use with Lesson 9-5, text pages 304–305.** Copyright © William H. Sadlier, Inc. All rights reserved.

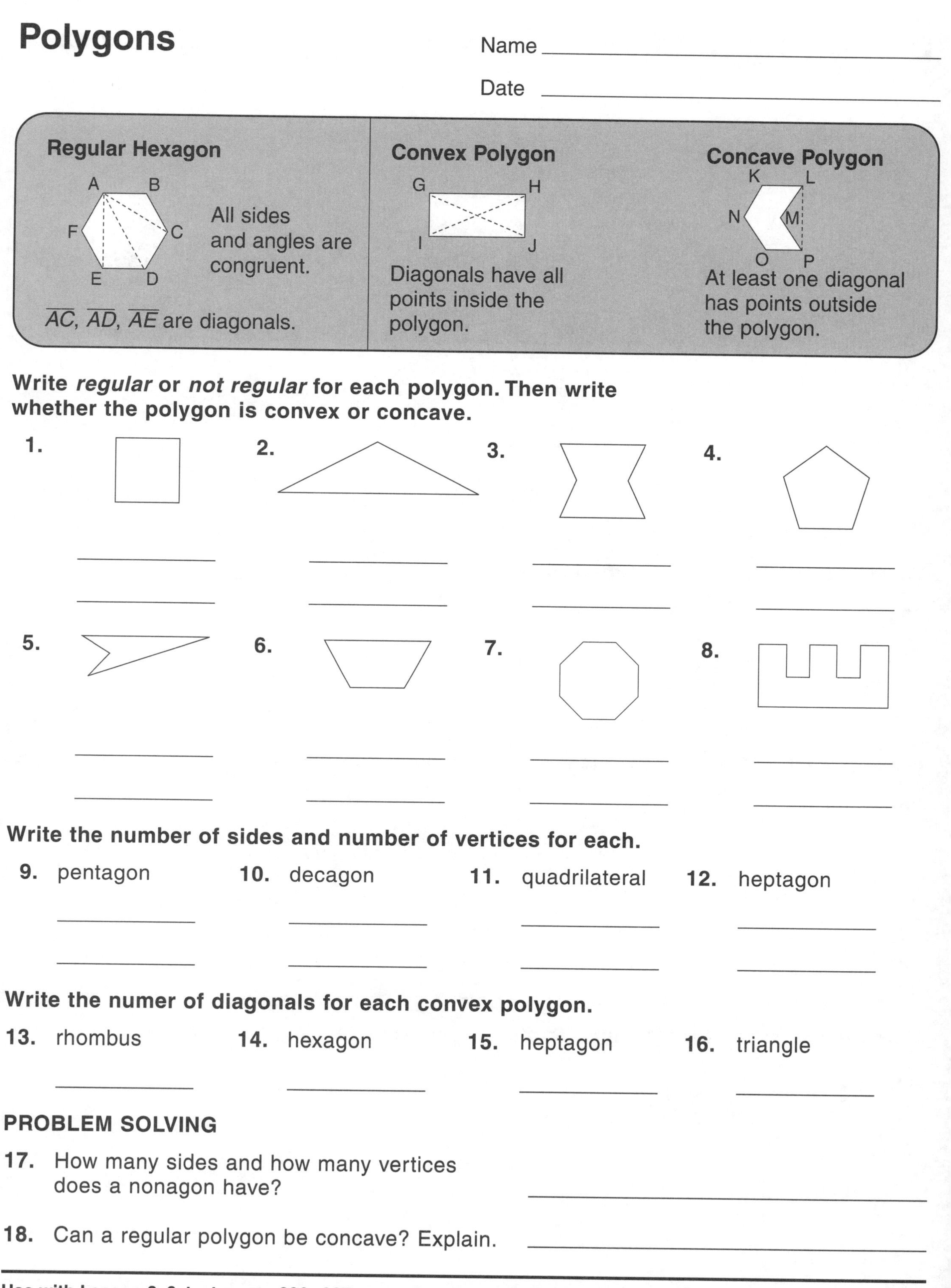

# Polygons

Name ______________________

Date ______________________

**Regular Hexagon**



All sides and angles are congruent.

$\overline{AC}$, $\overline{AD}$, $\overline{AE}$ are diagonals.

**Convex Polygon**



Diagonals have all points inside the polygon.

**Concave Polygon**



At least one diagonal has points outside the polygon.

**Write *regular* or *not regular* for each polygon. Then write whether the polygon is convex or concave.**

**1.** ______ ______

**2.** ______ ______

**3.** ______ ______

**4.** ______ ______

**5.** ______ ______

**6.** ______ ______

**7.** ______ ______

**8.** ______ ______

**Write the number of sides and number of vertices for each.**

**9.** pentagon ______ ______

**10.** decagon ______ ______

**11.** quadrilateral ______ ______

**12.** heptagon ______ ______

**Write the numer of diagonals for each convex polygon.**

**13.** rhombus ______

**14.** hexagon ______

**15.** heptagon ______

**16.** triangle ______

**PROBLEM SOLVING**

**17.** How many sides and how many vertices does a nonagon have? ______________

**18.** Can a regular polygon be concave? Explain. ______________


Copyright © William H. Sadlier, Inc. All rights reserved.

# Classifying Triangles

Name ______________________________

Date ______________________________

Triangles are classified by the length of their sides and by the measure of their angles.

| **scalene** | **isosceles** | **equilateral** | **acute** | **right** | **obtuse** |
|---|---|---|---|---|---|
| no sides congruent | 2 sides congruent | all sides congruent | 3 acute angles | 1 right angle | 1 obtuse angle |

**Label each triangle *equilateral, isosceles,* or *scalene.***

1. ____________ 2. ____________ 3. ____________ 4. ____________

**Label each triangle *acute, obtuse,* or *right.***

5. ____________ 6. ____________ 7. ____________ 8. ____________

**Classify each triangle by the measures of its sides and by the measures of its angles.**

9. 6 cm, 6 cm, 2 cm; 24°, 78°, 78° ____________

10. 3 m, 4 m, 5 m; 53°, 37° ____________

11. 3.8 m, 2.5 m, 2 m; 30°, 40°, 110° ____________

12. 3 in., 3 in., 3 in.; 60°, 60°, 60° ____________

**Find the measure of the missing angle in each triangle.**

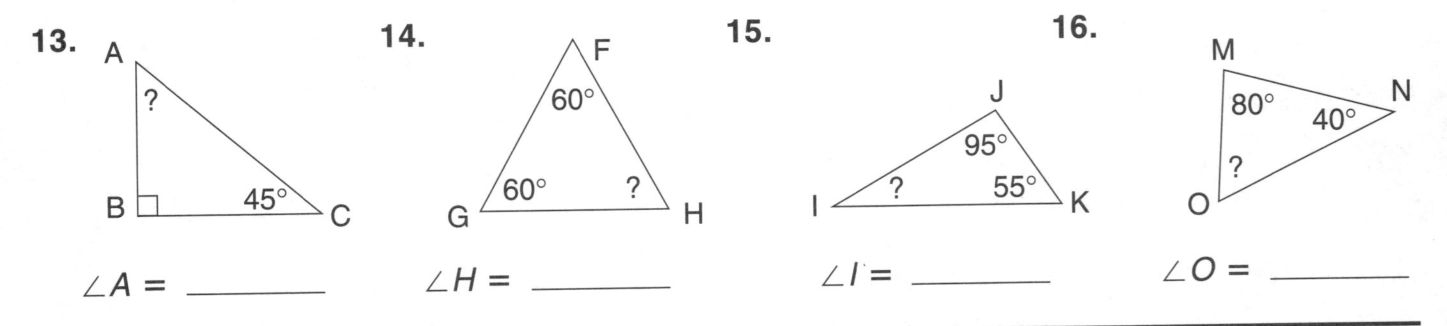


 Use with Lesson 9-7, text pages 308–309. Copyright © William H. Sadlier, Inc. All rights reserved.

# Classifying Quadrilaterals

Name ____________________

Date ____________________

parallelogram | rectangle | square | rhombus | trapezoid

**Write *True* or *False* for each statement.**

1. All rectangles are quadrilaterals. ______
2. All rectangles are parallelograms. ______
3. All parallelograms are rectangles. ______
4. Some rhombuses are squares. ______

**Complete the table. Write *Yes, No,* or *Sometimes* for each description.**

| | Description | Rhombus | Square | Rectangle | Parallelogram | Trapezoid |
|---|---|---|---|---|---|---|
| 5. | 4 right angles | | | | | |
| 6. | opposite sides parallel | | | | | |
| 7. | all sides congruent | | | | | |

**Find the measure of the fourth angle.**

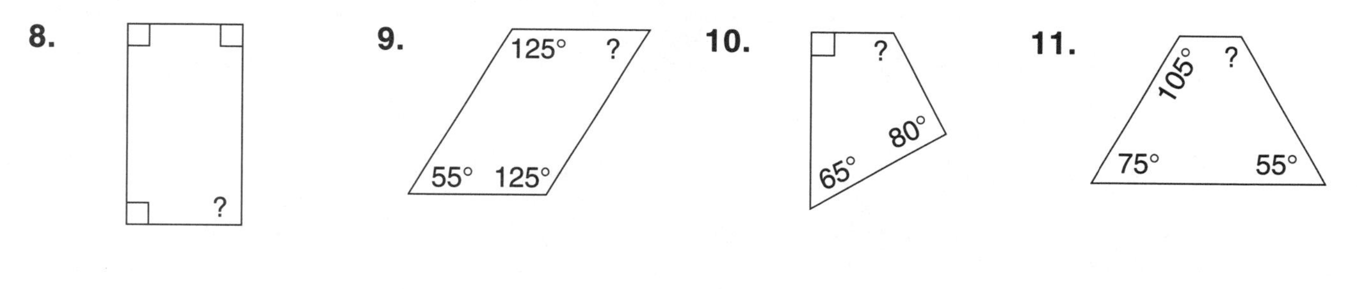


8. ________ 9. ________ 10. ________ 11. ________

**Construct the figure described.**

12. parallelogram *WXYZ*

________________________

13. rhombus *RSTU* with each side 2 cm long

________________________

 Copyright © William H. Sadler, Inc. All rights reserved. 

# Circles

Name ______________________

Date ______________________

**Circle: *M***



Center: *M*
Diameter: $\overline{XY}$
Radii: $\overline{XM}$, $\overline{MY}$, $\overline{MT}$
Central angle: $\angle YMT$, $\angle XMT$, $\angle XMY$
Chords: $\overline{RS}$, $\overline{XY}$

**Identify each for Circle A.**



1. the center of the circle ______________
2. two diameters ______________
3. five radii ______________
4. three chords ______________
5. five central angles ______________
6. six points on the circle ______________

**Write *True* or *False* for each statement.**

7. One endpoint of a radius of circle Q is point Q.

______________________

8. All chords pass through the center of a circle.

______________________

9. An arc is a curved path between two points on a circle.

______________________

10. A central angle of a circle has its vertex on the circle.

______________________

11. The radius of a circle is twice the length of its diameter.

______________________

12. All diameters form a straight angle.

______________________

**Use a compass to draw a circle with a radius of length 2 cm. Then draw and label each of the following:**

13. center: point *O*
14. radius: $\overline{OR}$
15. chord: $\overline{RS}$
16. central angle: $\angle ROT$
17. diameter: $\overline{XY}$
18. chord: $\overline{YZ}$

 **Use with Lesson 9-9, text pages 312–313.** Copyright © William H. Sadlier, Inc. All rights reserved.

# Classifying Space Figures

Name ______________________________

Date ______________________________

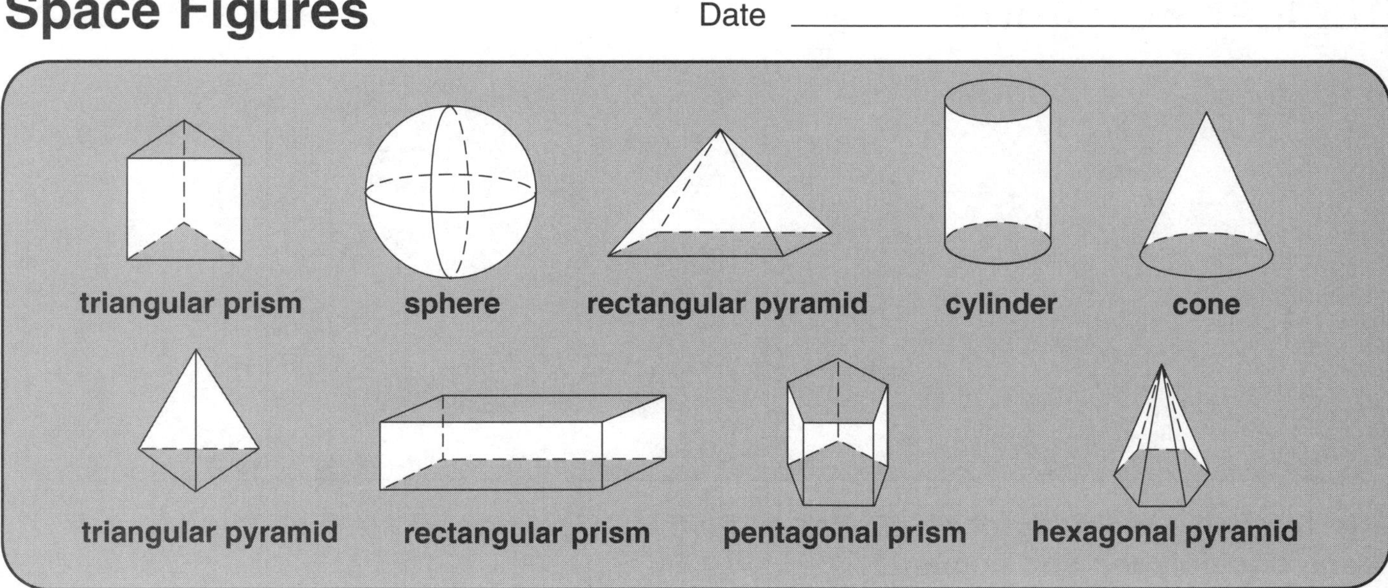


**Complete the table. Write the number of faces, vertices, and edges of each space figure.**

| | Space Figure | Faces | Vertices | Edges |
|---|---|---|---|---|
| **1.** | triangular prism | | | |
| **2.** | rectangular prism | | | |
| **3.** | pentagonal prism | | | |
| **4.** | triangular pyramid | | | |
| **5.** | rectangular pyramid | | | |
| **6.** | hexagonal pyramid | | | |

**Write which space figure(s) can have a base like the one shown.**

**7.** **8.** **9.** **10.**

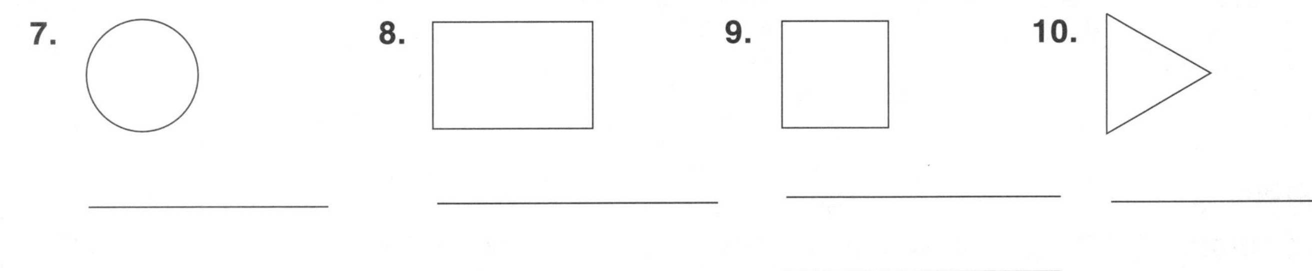

______________ ______________ ______________ ______________

______________ ______________ ______________ ______________

## PROBLEM SOLVING

**11.** A net of a space figure has 1 rectangular base and 4 triangular faces. Which space figure is it? How many edges does it have? How many vertices? ______________________________

**12.** Which of the space figures at the top of the page are *not* polyhedrons? ______________________________

 Copyright © William H. Sadlier, Inc. All rights reserved. 

# Congruent and Similar Polygons

Name ______________________

Date ______________________

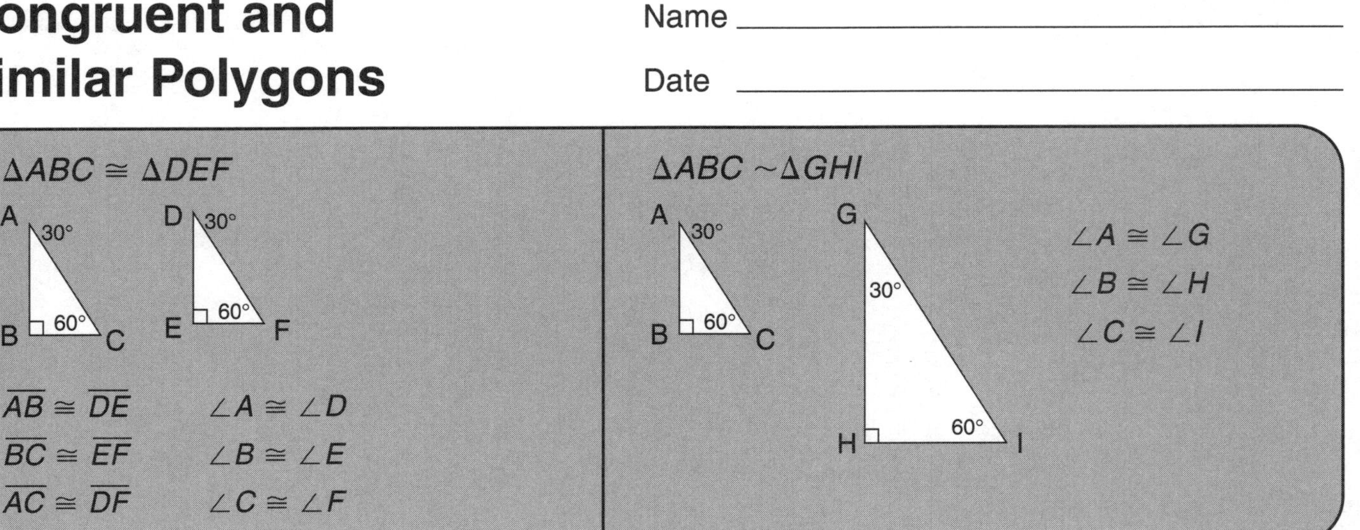


**Are the polygons *congruent, similar,* or *neither?* You may trace one polygon and compare.**

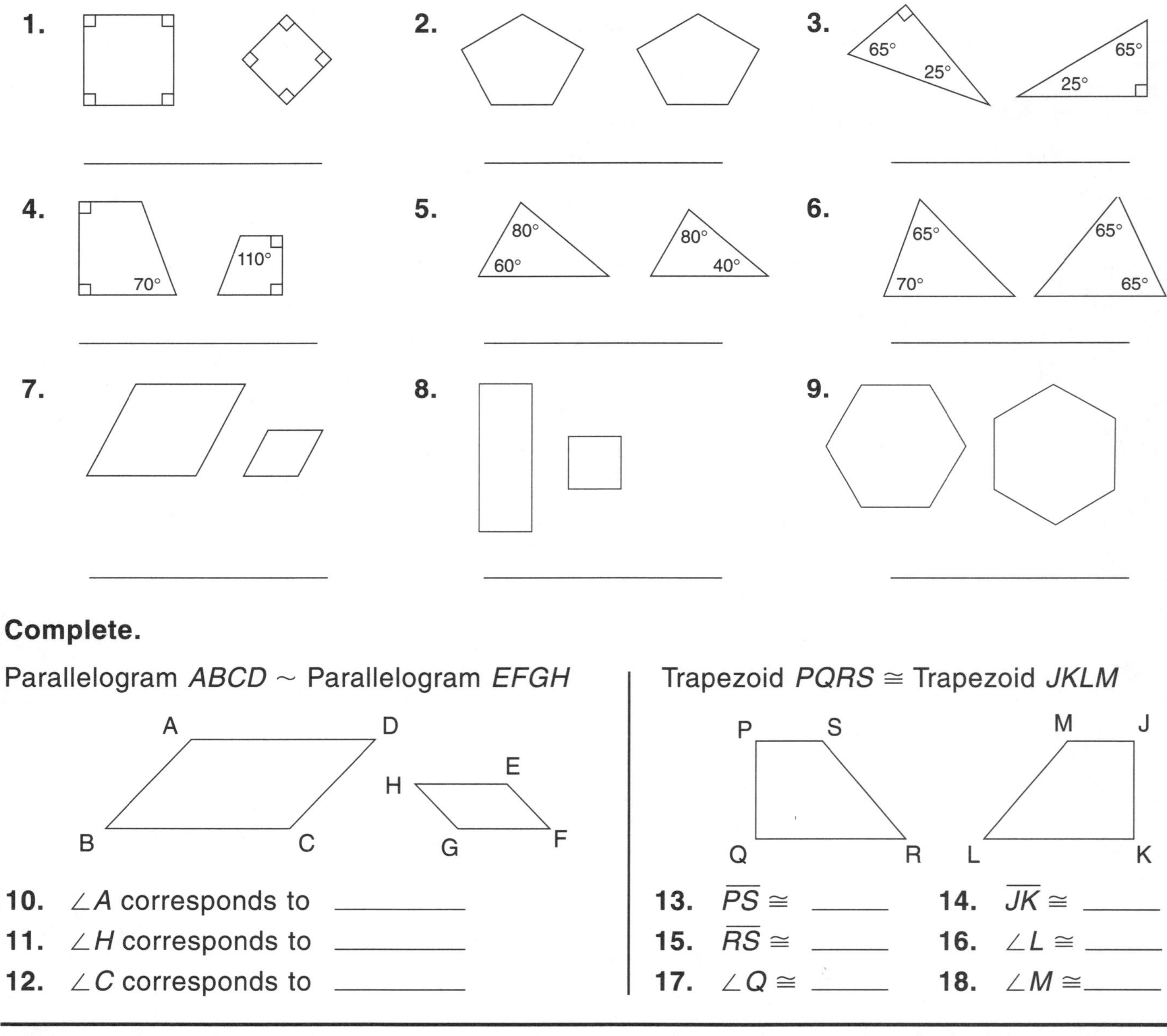


10. $\angle A$ corresponds to ________
11. $\angle H$ corresponds to ________
12. $\angle C$ corresponds to ________

13. $\overline{PS} \cong$ ______
14. $\overline{JK} \cong$ ______
15. $\overline{RS} \cong$ ______
16. $\angle L \cong$ ______
17. $\angle Q \cong$ ______
18. $\angle M \cong$ ______

 **Use with Lesson 9-11, text pages 316–317.** Copyright © William H. Sadlier, Inc. All rights reserved.

# Transformations

Name ______________________

Date ______________________

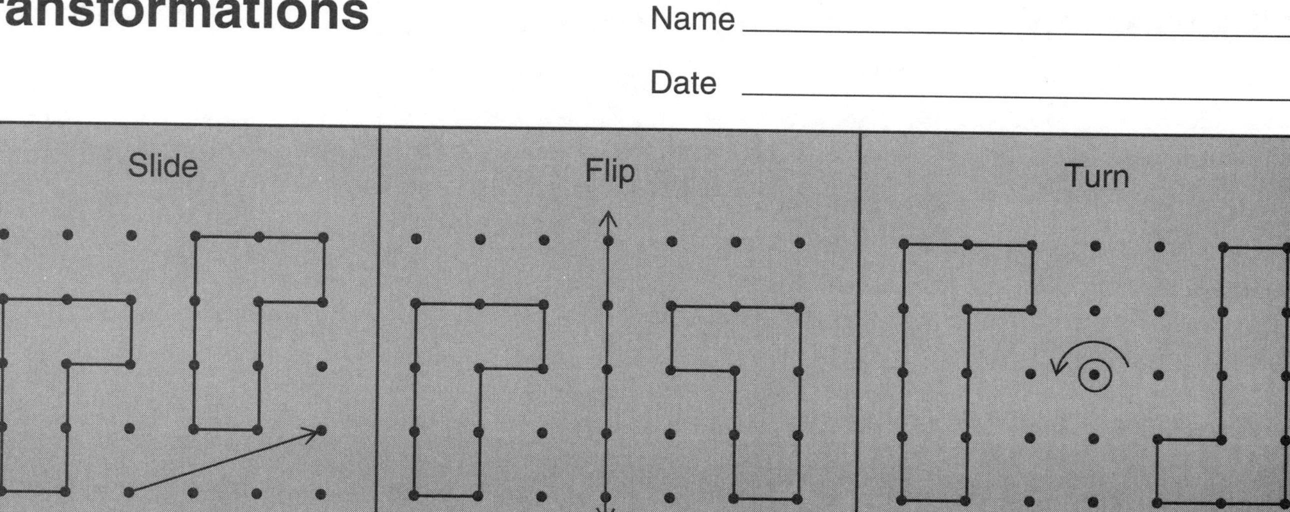


**Identify the transformation as a *slide*, *flip*, or *turn*. Then name the corresponding congruent sides and congruent angles in the transformation.**

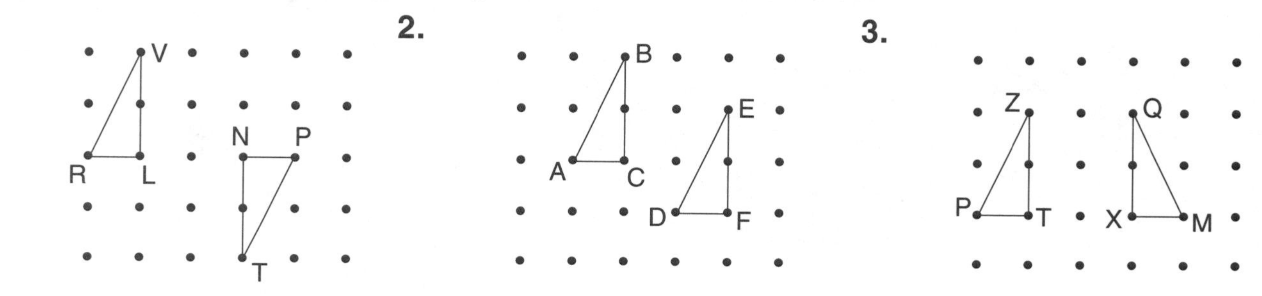


______________________ ______________________ ______________________

______________________ ______________________ ______________________

**Draw a slide, flip, or turn image of each figure.**

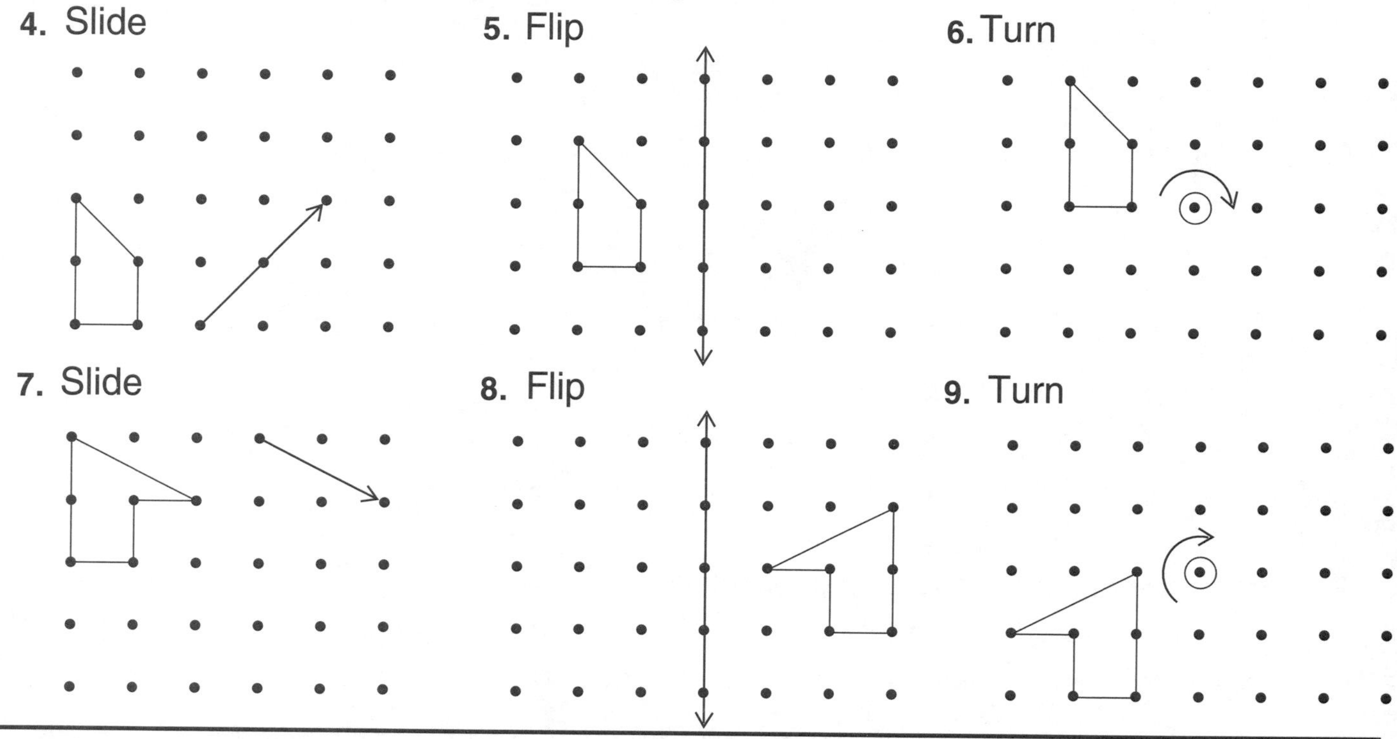


Use with Lesson 9-12, text pages 318–319.

Copyright © William H. Sadlier, Inc. All rights reserved.

# Tessellations

Name ________________________________

Date ________________________________

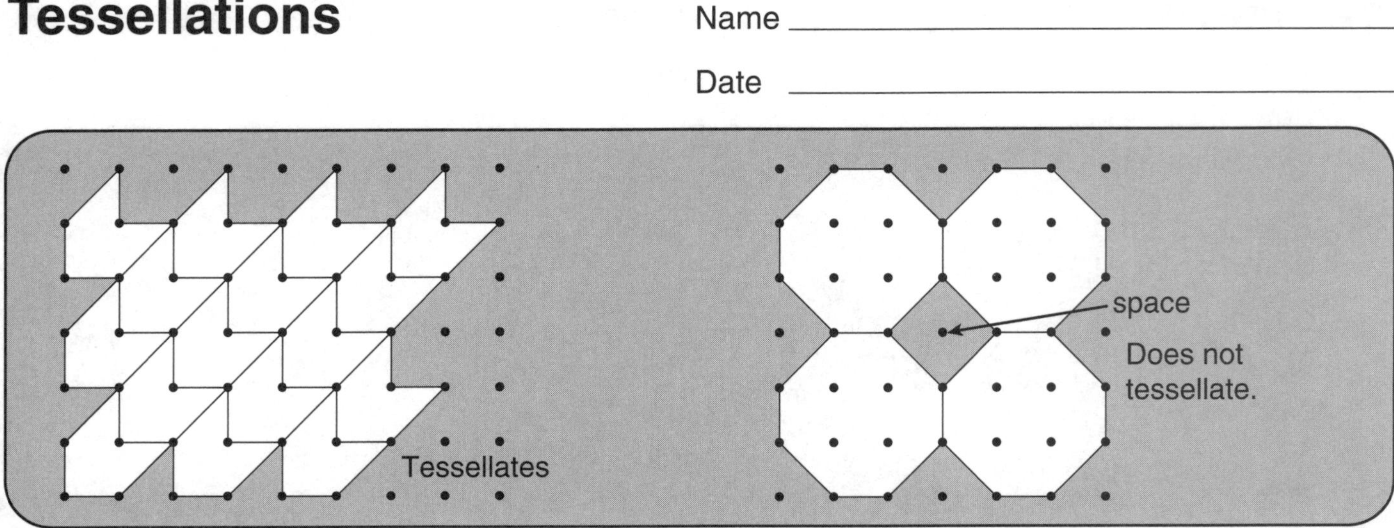


**Try to make a tessellation using each polygon. Does the polygon tessellate? Write *Yes* or *No.* Check students' tessellations**

**1.** ____________

**2.** ____________

**3.** ____________

**4.** ____________

**5.** Tessellate the plane using both squares and hexagons.

**6.** Tessellate the plane using both squares and trapezoids.

**7.** Circle the names of shapes that tessellate.

| squares | triangles | regular pentagons | regular hexagons |
|---|---|---|---|
| regular octagons | parallelograms | trapezoids | rectangles |

 Copyright © William H. Sadlier, Inc. All rights reserved.

# Problem-Solving Strategy: Logic/Analogies

Name ______________________

Date ______________________

A statement that tells the way in which two pairs of things are alike is an analogy. Choose the correct answer to make an analogy out of this incomplete statement.

Q is to Ɑ as F is to ? — First Pair, Second Pair, Possible Answers

Since Ɑ is upside down, look for an upside down F.

The answer is ᒷ.

**PROBLEM SOLVING Circle the letter of the correct choice for problems 1–6.**

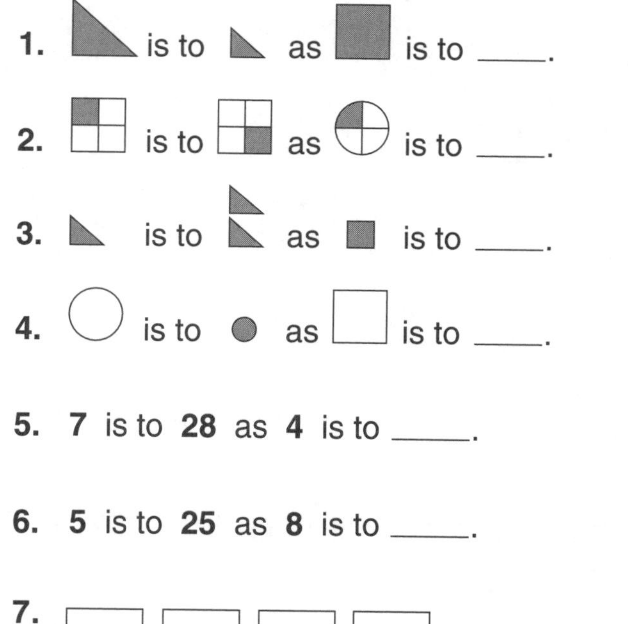

1. is to as is to ____. **a.** **b.** **c.** **d.**
2. is to as is to ____. **a.** **b.** **c.** **d.**
3. is to as is to ____. **a.** **b.** **c.** **d.**
4. is to as is to ____. **a.** **b.** **c.** **d.**
5. **7** is to **28** as **4** is to ____. **a.** **8** **b.** **16** **c.** **20** **d.** **2**
6. **5** is to **25** as **8** is to ____. **a.** **16** **b.** **32** **c.** **64** **d.** **80**

**7.**

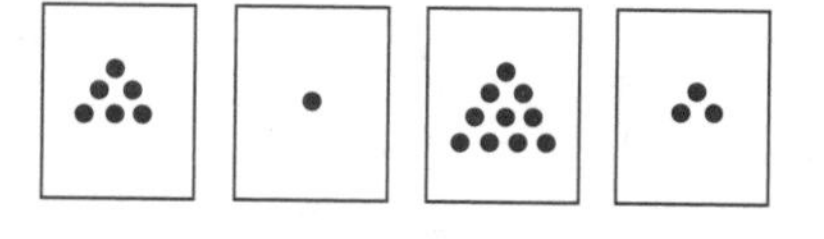

The cards can be rearranged in a row to form a pattern. Draw the arrangement, including a 5th card. Then describe the pattern.

______________________

**8.**

The cards can be rearranged in a row to form a pattern. Draw the arrangement, including a 5th card. Then describe the pattern.

______________________


Copyright © William H. Sadlier, Inc. All rights reserved.

# Measuring Metric Length

Name ______________________________

Date ______________________________

1 meter (m) = 1000 millimeters (mm)
= 100 centimeters (cm)
= 10 decimeters (dm)

1 kilometer (km) = 1000 meters

**Measure each line segment to the nearest centimeter and to the nearest millimeter.**

**1.** •——————•

**2.** •——————•

**3.** •———•

**4.** •——————•

**5.** •——————•

**Complete. Write *mm, cm, dm, m,* or *km*.**

**6.** The width of a postage stamp is about 2.5 ______.

**7.** The height of a flagpole is about 30 ______.

**8.** The distance between Norfolk and Newport News is about 35 ______.

**9.** The length of a key is about 45 ______.

**10.** The height of a full grown tree is about 400 ______.

**Draw each quadrilateral.**

**11.** square *ABCD* with *AB* = 19 mm

**12.** paralleogram *EFGH* with *EF* = 2.7 cm and *EH* = 1.3 cm

**PROBLEM SOLVING**

**13.** How many 1-cm pieces of metal can be cut from a rod that is 2.3 m long? ______________________

**14.** Ribbon A and ribbon B are the same length. Tom says that one is 24.6 cm long and the other is 246 mm long. Can he be correct? Explain. ______________________

______________________


Copyright © William H. Sadlier, Inc. All rights reserved.

# Measuring Metric Capacity and Mass

Name ______________________

Date ______________________

| kiloliter (kL) | hectoliter (hL) | dekaliter (daL) | liter (L) | deciliter (dL) | centiliter (cL) | milliliter (mL) |
|---|---|---|---|---|---|---|
| 1000 L | 100 L | 10 L | 1 L | 0.1 L | 0.01 L | 0.001 L |

| kilogram (kg) | hectogram (hg) | dekagram (dag) | gram (g) | decigram (dg) | centigram (cg) | milligram (mg) |
|---|---|---|---|---|---|---|
| 1000 g | 100 g | 10 g | 1 g | 0.1 g | 0.01 g | 0.001 g |

**Compare. Write <, =, or >.**

1. 800 L ____ 8 daL
2. 2000 kg ____ 2 g
3. 4.5 L ____ 4500 mL
4. 26 g ____ 0.26 dg
5. 6.7 L ____ 67 cL
6. 5300 g ____ 5.3 kg
7. 9000 mL ____ 90 L
8. 850 mg ____ 8.5 g
9. 7650 cL ____ 7.65 L
10. 2800 g ____ 2.8 kg
11. 7.8 mL ____ 0.78 L
12. 630 hg ____ 63 g
13. 1.7 hL ____ 17 L
14. 350 dg ____ 35 g
15. 16 000 mL ____ 160 L

**PROBLEM SOLVING**

16. Alonzo needs 1.65 L of distilled water for an experiment. How many times should he fill a container that holds 550 mL with the water to be sure he has enough? ______________________

17. Mr. Wong has a box that contains 1 kg of salt. Is there enough salt in the box for him to give 42 g to each of his 25 science students? ______________________

18. A bottle holds 750 mL of liquid. How many liters of liquid do 8 bottles hold? ______________________

19. A dictionary has a mass of 5.85 kg. A large telephone directory has a mass of 5625 g. Which has the greater mass? ______________________

20. How many liters are there in 72 695 mL? ______________________


Copyright © William H. Sadlier, Inc. All rights reserved.

# Renaming Metric Units

Name ______________________

Date ______________________

| Multiply to rename larger units as smaller units. | Divide to rename smaller units as larger units. |
| --- | --- |
| 8.5 km = ? m | 6200 g = ? hg |
| 8.5 km = (8.5 × 1000) m = 8500 m | 6200 g = (6200 ÷ 100) hg = 62 hg |
| or | or |
| 8.5 km = 8.500 m = 8500 m | 6200 g = 6200 hg = 62 hg |

**Complete.**

**1.** 8 m = ______ cm
**2.** 80 cm = ______ m
**3.** 3 cm = ______ mm
**4.** 25 cm = ______ mm
**5.** 3560 mm = ______ m
**6.** 4.2 m = ______ cm
**7.** 5 m = ______ mm
**8.** 4000 m = ______ km
**9.** 5.2 m = ______ mm
**10.** 94.2 L = ______ mL
**11.** 0.006 kL = ______ L
**12.** 7.5 cL = ______ mL
**13.** 4.375 L = ______ mL
**14.** 18 400 L = ______ kL
**15.** 0.025 kL = ______ mL
**16.** 6 kg = ______ g
**17.** 1.5 g = ______ mg
**18.** 73 mg = ______ g
**19.** 38 mg = ______ g
**20.** 0.025 kg = ______ mg
**21.** 5.7 mg = ______ cg
**22.** 8 g = ______ kg
**23.** 35 hg = ______ dag
**24.** 57 000 mg = ______ g

**Compare. Write <, =, or >.**

**25.** 3 L ____ 300 cL
**26.** 35.5 cL ____ 3.55 L
**27.** 650 L ____ 0.065 kL
**28.** 3800 mL ____ 0.38 L
**29.** 1500 mL ____ 15 L
**30.** 299 L ____ 0.299 kL
**31.** 0.6 kg ____ 6000 g
**32.** 20 000 mg ____ 0.02 kg
**33.** 0.75 kg ____ 75 g
**34.** 4 kg ____ 4000 mg
**35.** 0.68 g ____ 6800 mg
**36.** 0.52 kg ____ 520 g
**37.** 8.6 km ____ 860 m
**38.** 250 mm ____ 0.25 m
**39.** 8.33 m ____ 83 000 mm
**40.** 0.8 m ____ 8000 mm
**41.** 1400 mm ____ 0.14 m
**42.** 3.8 cm ____ 38 mm
**43.** 15 dm ____ 1.5 m
**44.** 23 dm ____ 2300 cm
**45.** 4.6 cm ____ 0.46 dm

**PROBLEM SOLVING**

**46.** An audio cassette is 1 dm long. How many meters long is a rack in a music store that holds 36 audio cassettes laid lengthwise? ______________

 **Use with Lesson 10-4, text pages 338–339.** Copyright © William H. Sadlier, Inc. All rights reserved.

# Relating Metric Units

Name ______________________

Date ______________________

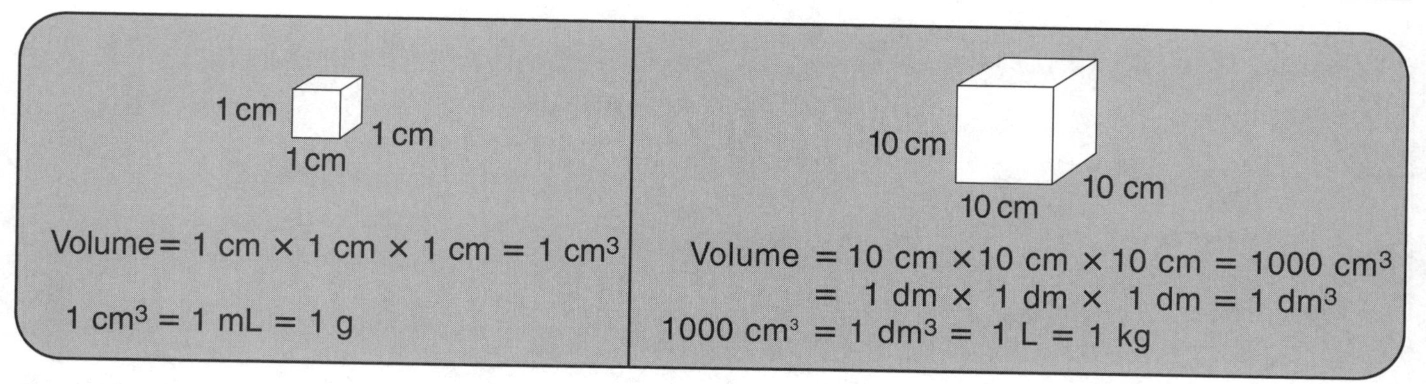


**Complete each chart.**

| | Cube | Capacity | Mass |
|---|---|---|---|
| **1.** | 3 cm³ | 3 mL | |
| **2.** | 10 cm³ | | |
| **3.** | | 2 mL | |
| **4.** | | | 8 g |
| **5.** | 4.5 dm³ | | |
| **6.** | 5800 cm³ | | |
| **7.** | | 3.7 mL | |
| **8.** | | | 21.5 g |

| | Cube | Capacity | Mass |
|---|---|---|---|
| **9.** | 2 dm³ | 2 L | |
| **10.** | | 8 L | |
| **11.** | 2.1 dm³ | | |
| **12.** | | | 20 kg |
| **13.** | 7400 cm³ | | |
| **14.** | | | 8.5 kg |
| **15.** | 67.5 cm³ | | |
| **16.** | | 9.6 L | |

**Measure the edge of each cube to the nearest 0.5 cm. Then find its capacity and mass.**

**17.**

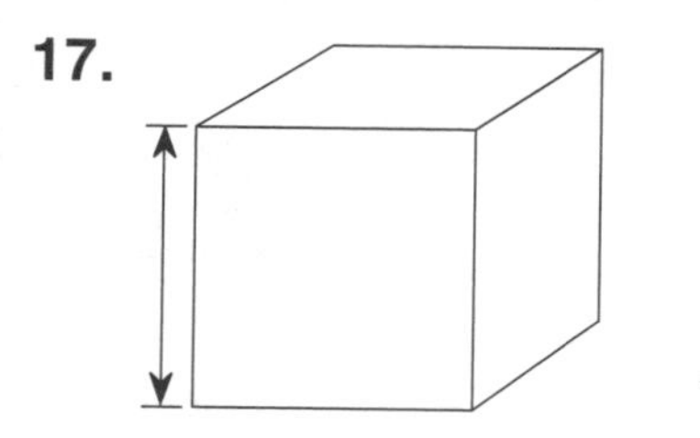

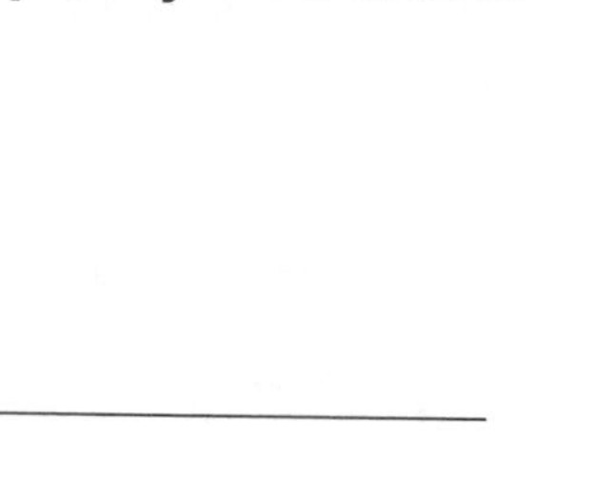

______________________

**18.**

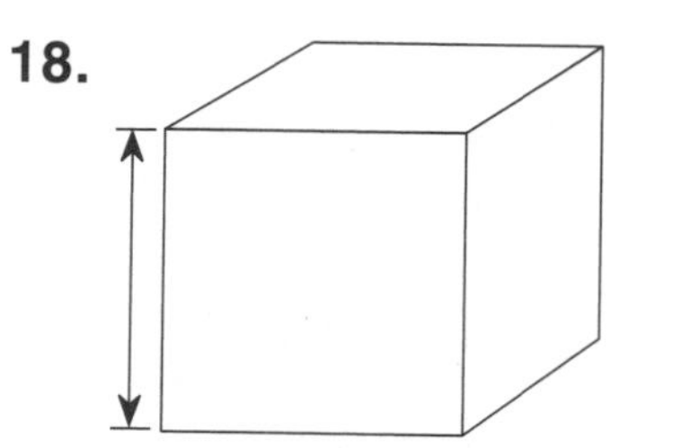

______________________

## PROBLEM SOLVING

**19.** A wading pool holds 1200 kg of water. What is its capacity? ______________________

**20.** Each edge of a cube measures 20 cm. Loren fills the cube with water. What is the mass of the water in the cube in grams? in kilograms? ______________________


Copyright © William H. Sadlier, Inc. All rights reserved.

# Measuring Customary Length, Capacity, and Weight

Name ______________________

Date ______________________

| **Multiply to rename larger units as smaller units:** | **Divide to rename smaller units as larger units.** |
|---|---|
| $8\frac{1}{3}$ yd = ? ft | 54 oz = ? lb |
| $8\frac{1}{3}$ yd = $(8\frac{1}{3} \times 3)$ ft | 54 oz = (54 ÷ 16) lb |
| $= \frac{25}{3} \times \frac{3}{1}$ ft = 25 ft | = 3 lb 6 oz |
| | $= 3\frac{3}{8}$ lb ($\frac{6}{16}$ or $\frac{3}{8}$ lb) |

**Measure each to the nearest in., $\frac{1}{2}$ in., $\frac{1}{4}$ in., $\frac{1}{8}$ in., and $\frac{1}{16}$ in.**

**1.** ______________________

**2.** ______________________

**3.** ______________________

**Complete.**

**4.** 3 ft = _____ in.

**5.** 12 pt = _____ qt

**6.** 5 qt = _____ pt

**7.** 24 yd = _____ ft

**8.** 7 pt = _____ c

**9.** $1\frac{1}{2}$ gal = _____ qt

**10.** $3\frac{1}{4}$ mi = _____ ft

**11.** 4 lb 15 oz = _____ oz

**12.** 4.5 T = _____ lb

**13.** 80 ft = _____ yd

**14.** 8 oz = _____ lb

**15.** $6\frac{1}{8}$ pt = _____ fl oz

**16.** 10,560 ft = _____ mi

**17.** $4\frac{3}{4}$ lb = _____ oz

**18.** 2400 lb = _____ T _____ lb

**Compare. Write $<$, $=$, or $>$.**

**19.** 52 ft _____ 18 yd

**20.** 4 c 7 fl oz _____ 40 fl oz

**21.** 256 oz _____ 15 lb 4 oz

**22.** 5 gal 2 qt _____ 24 qt

**23.** 12 T 850 lb _____ 25,000 lb

**24.** 224 in. _____ 18 ft 8 in.

**25.** 72 in. _____ 6 ft

**26.** 100 qt _____ 25 gal

**27.** 36 lb _____ 524 oz

## PROBLEM SOLVING

**28.** A length of clothesline is $75\frac{1}{2}$ ft long. How would you report its length in inches? in yards? ______________________

**29.** How many pints are equal to $3\frac{3}{4}$ gal? ______________________

 Copyright © William H. Sadlier, Inc. All rights reserved.

# Computing Customary Units

Name ______________________

Date ______________________

| | |
|---|---|
| 2 ft 9 in. <br> +3 ft 8 in. <br> 5 ft 17 in. = 6 ft 5 in. <br> 17 in. → 1 ft 5 in. | (4 + 1) <br> ~~5~~ gal 1 qt → 4 gal 5 qt <br> − 3 gal 3 qt → − 3 gal 3 qt <br> 1 gal 2 qt |
| 3 ft 7 in. <br> × 3 <br> 9 ft 21 in. = 10 ft 9 in. <br> 21 in. → 1 ft 9 in. | 4 lb 6 oz ÷ 2 = ? <br> 64 oz + 6 oz ÷ 2 = ? <br> 70 oz ÷ 2 = 35 oz = 2 lb 3 oz |

**Add.**

1. 8 ft 9 in. + 2 ft 6 in.
2. 3 gal 2 qt + 2 gal 3 qt
3. 10 yd 1 ft + 8 yd 2 ft
4. 5 pt 2 c + 8 pt 3 c

**Subtract.**

5. 7 ft 9 in. − 3 ft 10 in.
6. 8 yd 4 in. − 5 yd 8 in.
7. 6 gal 1 qt − 3 gal 3 qt
8. 12 pt 1 c − 8 pt 2 c

**Multiply.**

9. 4 ft 3 in. × 3
10. 7 yd 4 ft × 6
11. 3 mi 26 yd × 8
12. 6 qt 1 pt × 5

**Divide.**

13. 3 gal 3 qt ÷ 5 = ______
14. 6 yd 2 ft ÷ 4 = ______
15. 3 lb 6 oz ÷ 3 = ______
16. 6 ft 8 in. ÷ 4 = ______
17. 12 qt 1 pt ÷ 5 = ______
18. 2 mi 500 ft ÷ 4 = ______

**PROBLEM SOLVING**

19. Find the combined length of two sticks that are 4 yd 2 ft and 6 yd 2 ft long. ______

20. David's fishing pole is 7 ft 4 in. long. Paul's is 9 ft 2 in. long. How much longer is Paul's fishing pole? ______


Copyright © William H. Sadlier, Inc. All rights reserved.

# Using Perimeter

Name ______________________________

Date ______________________________

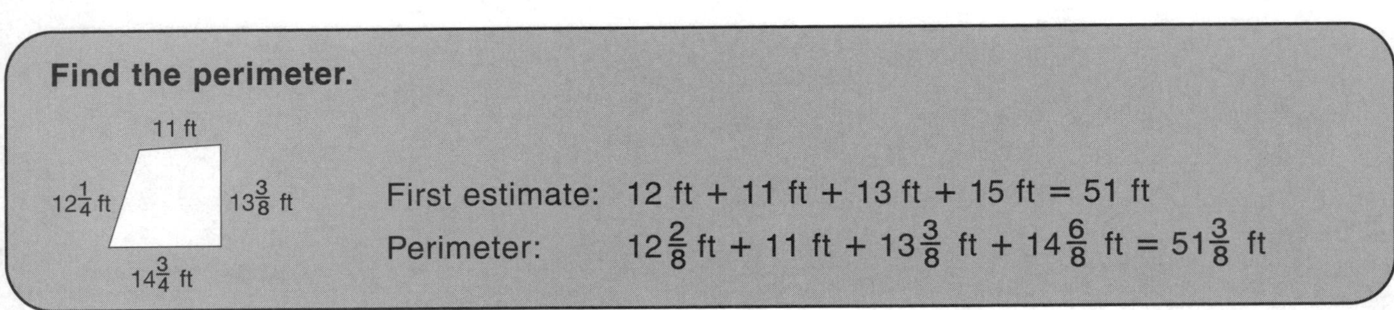


**Draw a diagram and then solve each problem.**
**Use a perimeter formula when you can.**

1. How many meters of fencing are needed to enclose a rectangular garden that is 5.3 m long and 3.7 m wide?

   ______________________________

2. How many feet of wallpaper border is needed to go around the walls of a rectangular room that measures $3\frac{1}{2}$ yd by 4 yd?

   ______________________________

3. Shirelle is taping crepe paper to the edges of a square table that is 121.9 cm on a side. How much crepe paper does she need?

   ______________________________

4. What is the perimeter of a quadrilateral with sides that measure $14\frac{1}{3}$ ft, $12\frac{3}{4}$ ft, 9 ft, and 10 ft?

   ______________________________

5. Ms. Reyes will put a border of cloth tape around a rectangular quilt that measures 72 in. by 85 in. If she allows for an extra 2 in. of tape to turn each corner, how many full yards of the tape should she buy?

   ______________________________

6. Satin ribbon costs $3.65 per yard. Alex will glue two stripes of ribbon around a square box that is 18 in. on a side. How much will Alex pay for the ribbon?

   ______________________________

7. What is the perimeter of an equilateral triangle with sides that measure 18.4 cm?

   ______________________________

8. What is the perimeter of a rhombus that measures 35.6 mm along one side?

   ______________________________

9. Enrico has 130 cm of gold trim. Will this be enough to make a border for a pennant with sides of 20.4 cm, 57.8 cm, and 57.8 cm?

   ______________________________

10. What is the perimeter of a rectangular football field, including the end zones, if it measures 120 yd by 55 yd?

    ______________________________

11. One side of a regular hexagon measures 10.6 cm. What is its perimeter? ______________________________


Copyright © William H. Sadlier, Inc. All rights reserved.

# Area of Rectangles and Squares

Name ______________________

Date ______________________

$1\frac{1}{4}$ yd

$A = s \times s$ or $A = s^2$

$A = 1\frac{1}{4}\text{ yd} \times 1\frac{1}{4}\text{ yd}$

$A = 1\frac{9}{16}\text{ yd}^2$

5.8 km

12.5 km

$A = \ell \times w$

$A = 12.5\text{ km} \times 5.8\text{ km}$

$A = 72.5\text{ km}^2$

**Use formulas to find the areas. Estimate to help you.**

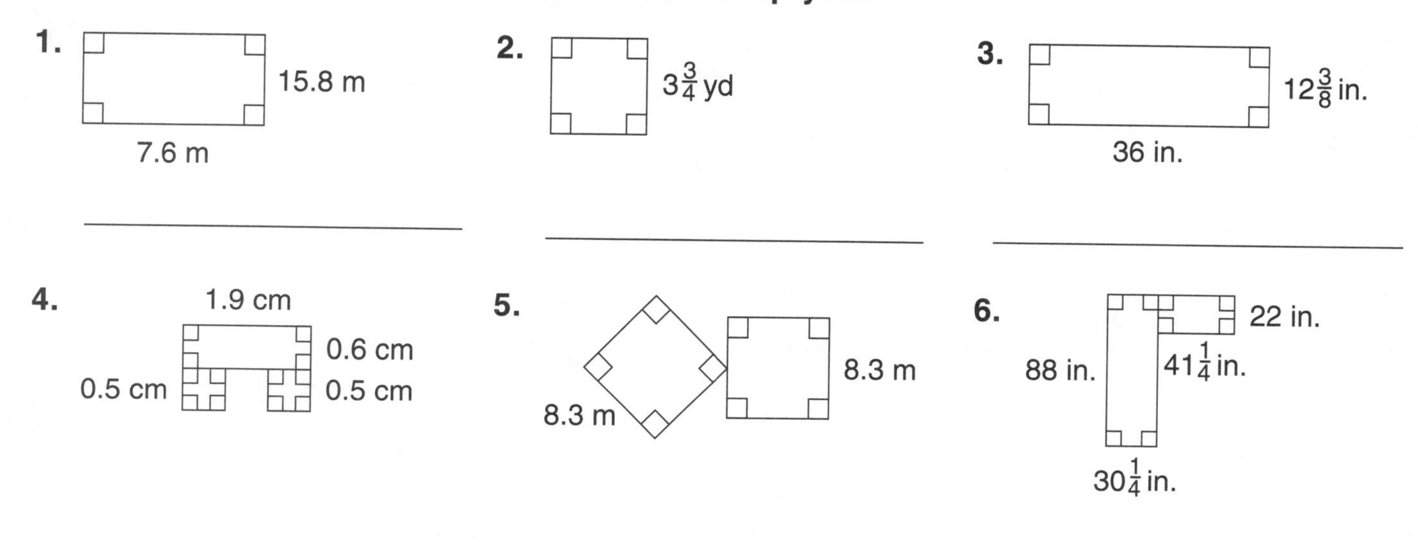


**PROBLEM SOLVING**

7. A baseball diamond is really a square that measures 90 feet along each base path. What is the area of a baseball diamond?

8. A tennis court is a rectangle that is 23.4 m long and 8.1 m wide. What is the area of a tennis court?

9. How many square yards of carpeting are needed to cover the floor of a rectangular room that is 21 ft long and 12 ft wide?

10. Floor tile is sold in 12-in. squares. How many tiles would you buy to cover the floor of a rectangular hallway that is 3 yd long and $1\frac{1}{3}$ yd wide?

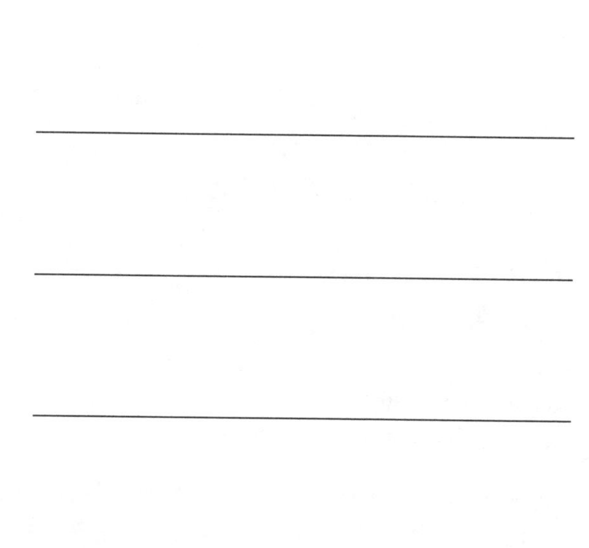

 Copyright © William H. Sadlier, Inc. All rights reserved. 

# Discovering Perimeter and Area

Name ______________________

Date ______________________

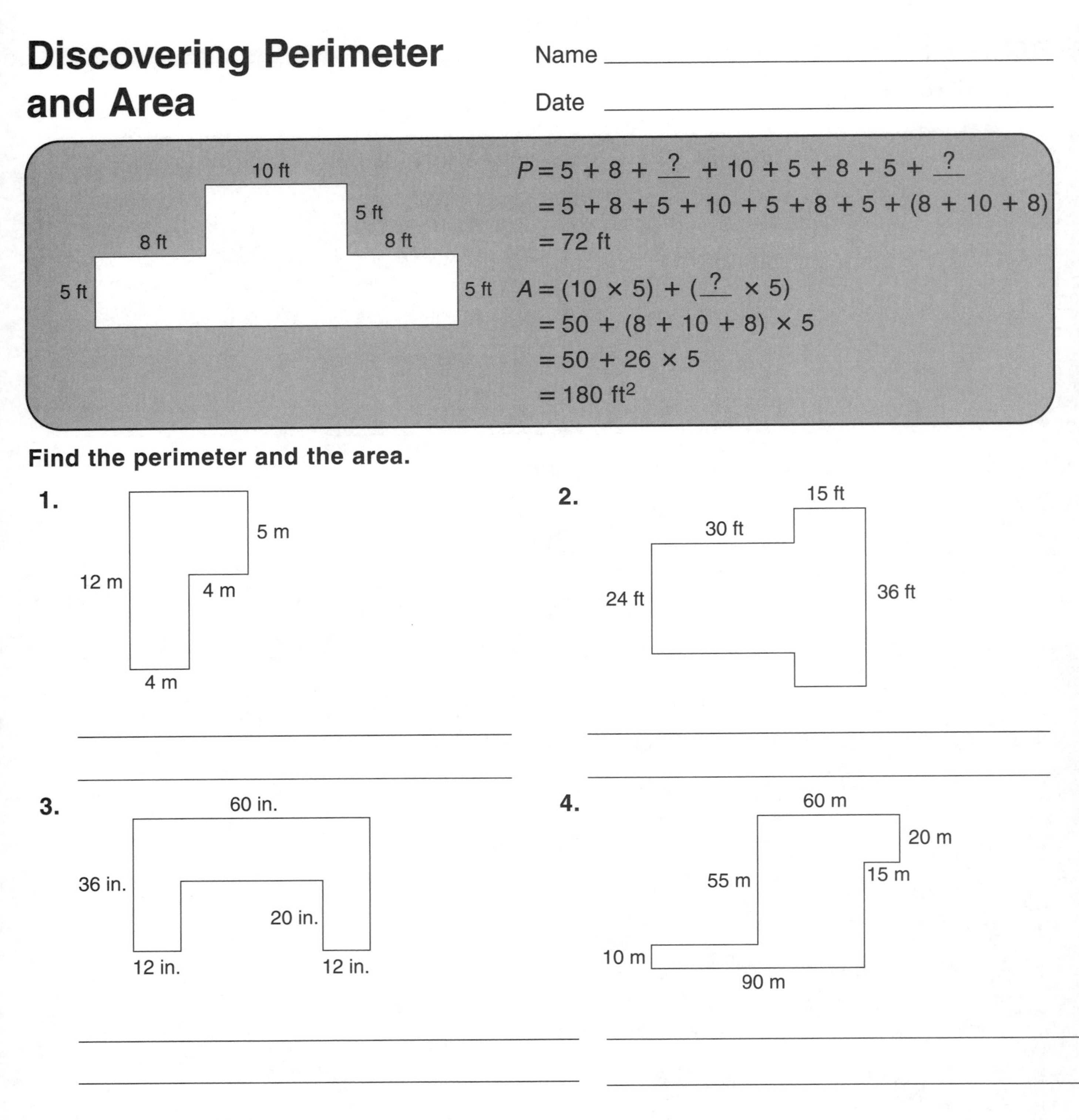


$P = 5 + 8 + \underline{?} + 10 + 5 + 8 + 5 + \underline{?}$
$= 5 + 8 + 5 + 10 + 5 + 8 + 5 + (8 + 10 + 8)$
$= 72$ ft

$A = (10 \times 5) + (\underline{?} \times 5)$
$= 50 + (8 + 10 + 8) \times 5$
$= 50 + 26 \times 5$
$= 180$ ft$^2$

**Find the perimeter and the area.**

**1.**


______________________

______________________

**2.**


______________________

______________________

**3.**


______________________

______________________

**4.**


______________________

______________________

______________________

**PROBLEM SOLVING**

5. Each floor tile is a 1-ft square and costs $9.95. How much would it cost to tile a floor with the same shape and dimensions as the figure in exercise 2? ______________________

6. How much would it cost to put fencing around a garden with the same shape and dimensions as the figure in exercise 1 if fencing costs $25.50 per meter? ______________________

 Copyright © William H. Sadlier, Inc. All rights reserved.

# Area of Triangles and Parallelograms

Name ______________________

Date ______________________

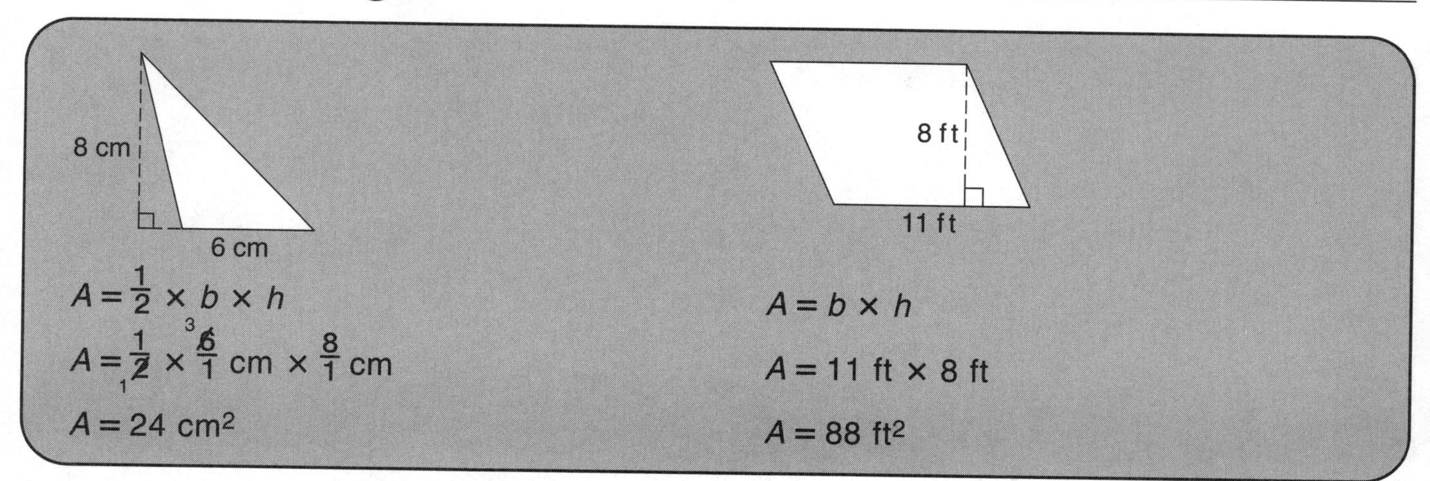


$A = \frac{1}{2} \times b \times h$

$A = \frac{1}{2} \times \frac{6}{1} \text{ cm} \times \frac{8}{1} \text{ cm}$

$A = 24 \text{ cm}^2$

$A = b \times h$

$A = 11 \text{ ft} \times 8 \text{ ft}$

$A = 88 \text{ ft}^2$

**Find the area of each triangle.**

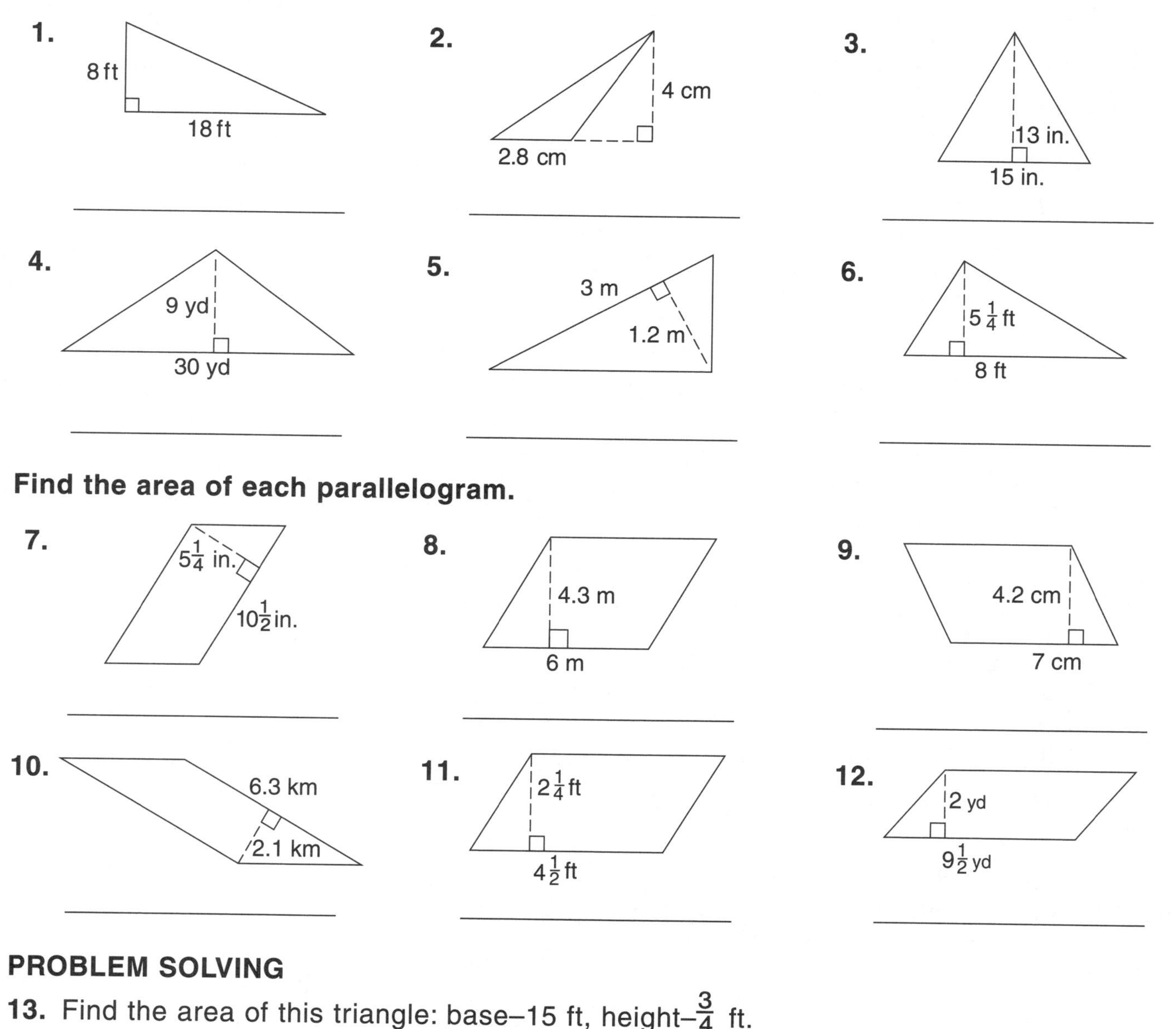


**PROBLEM SOLVING**

**13.** Find the area of this triangle: base–15 ft, height–$\frac{3}{4}$ ft. ______________

 Copyright © William H. Sadlier, Inc. All rights reserved. 

# Surface Area

Name ______________________________

Date ______________________________

Find the surface area of a rectangular prism.

2 cm, 4 cm, 3 cm

Area of bottom $= 4 \text{ cm} \times 3 \text{ cm} = 12 \text{ cm}^2$
Area of top $= 4 \text{ cm} \times 3 \text{ cm} = 12 \text{ cm}^2$
Area of side $= 2 \text{ cm} \times 3 \text{ cm} = 6 \text{ cm}^2$
Area of side $= 2 \text{ cm} \times 3 \text{ cm} = 6 \text{ cm}^2$
Area of front $= 4 \text{ cm} \times 2 \text{ cm} = 8 \text{ cm}^2$
Area of back $= 4 \text{ cm} \times 2 \text{ cm} = 8 \text{ cm}^2$
Surface Area (*S. A.*) $= 52 \text{ cm}^2$

Find the surface area of a cube.

3 in.

Area of one face $= 3 \text{ in.} \times 3 \text{ in.}$
$= 9 \text{ in.}^2$
*S.A.* $= 6 \times 9 \text{ in.}^2 = 54 \text{ in.}^2$

**Use the nets to find the surface area.**

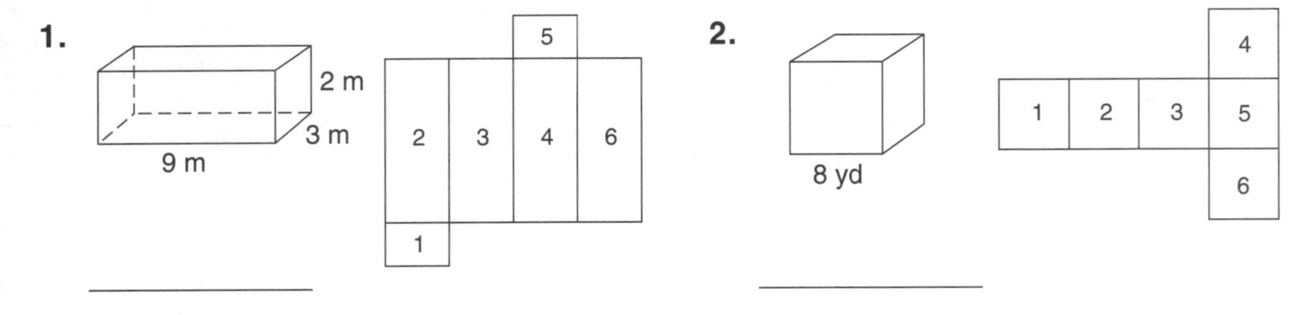


1. ______________ 2. ______________

**Find the surface area of each space figure.**

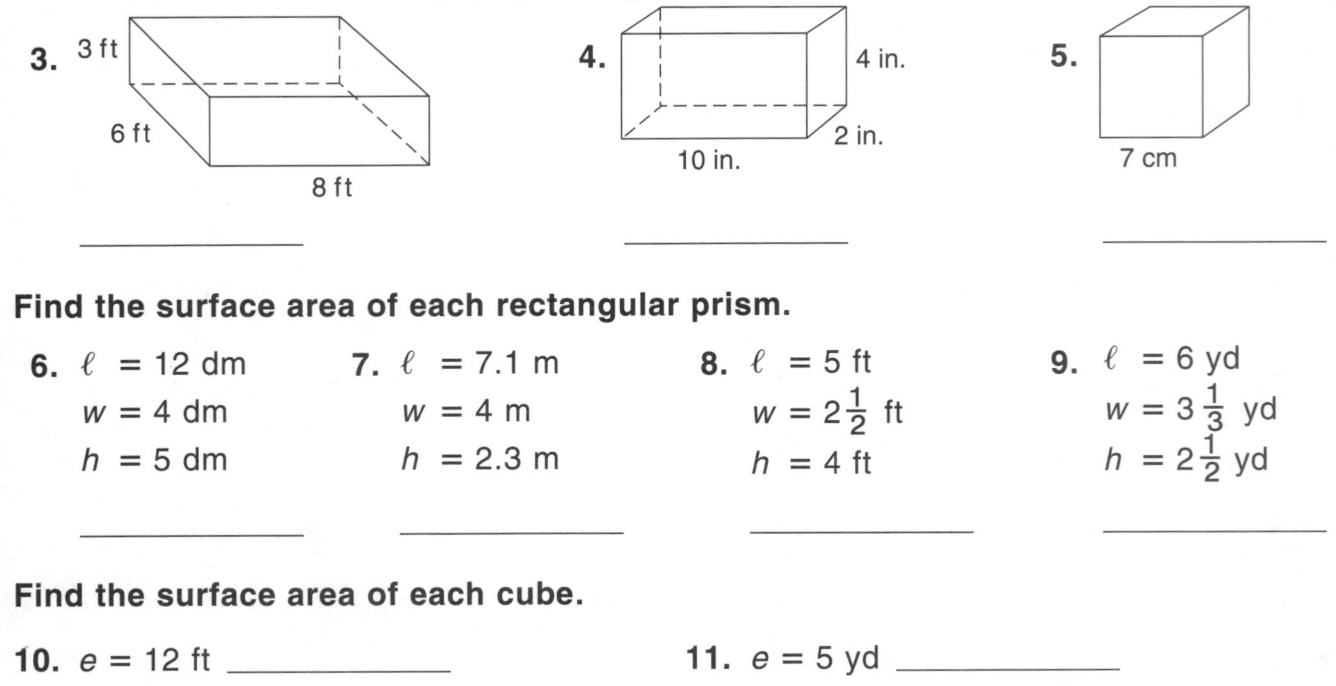


3. ______________ 4. ______________ 5. ______________

**Find the surface area of each rectangular prism.**

6. $\ell = 12$ dm, $w = 4$ dm, $h = 5$ dm ______________

7. $\ell = 7.1$ m, $w = 4$ m, $h = 2.3$ m ______________

8. $\ell = 5$ ft, $w = 2\frac{1}{2}$ ft, $h = 4$ ft ______________

9. $\ell = 6$ yd, $w = 3\frac{1}{3}$ yd, $h = 2\frac{1}{2}$ yd ______________

**Find the surface area of each cube.**

10. $e = 12$ ft ______________

11. $e = 5$ yd ______________

12. $e = 3.2$ m ______________

13. $e = 0.5$ dm ______________


Copyright © William H. Sadlier, Inc. All rights reserved.

# Circumference

Name ______________________________

Date ______________________________

| $C = \pi \times d$ (diameter 10.7 cm, $C = \underline{?}$) | $C = 2 \times \pi \times r$ (radius 12 ft, $C = \underline{?}$) |
|---|---|
| $C \approx 3.14 \times 10.7$ | $C \approx 2 \times 3.14 \times 12$ |
| $C \approx 33.598$ cm | $C \approx 75.36$ ft |

**Find the circumference of each circle.**

1. 18 in. ____________
2. 5.5 m ____________
3. 9 yd ____________
4. 4.7 cm ____________
5. 2.6 m ____________
6. 7.5 in. ____________
7. 11.1 cm ____________
8. 6 yd ____________

**Estimate the circumference. Use 3 for $\pi$.**

9. $d = 4.2$ cm ____________
10. $r = 6\frac{3}{4}$ mi ____________
11. $d = 15$ ft ____________
12. $r = 12.5$ m ____________

**Find the circumference. Use 3.14 or $\frac{22}{7}$ for $\pi$.**

13. $r = 1\frac{1}{6}$ in. ____________
14. $d = 10.5$ m ____________
15. $d = 21$ yd ____________
16. $r = 8.5$ dm ____________
17. $d = 3\frac{1}{2}$ ft ____________
18. $r = 2\frac{1}{3}$ mi ____________
19. $r = 30.2$ cm ____________
20. $d = 28$ mm ____________

**PROBLEM SOLVING**

21. The diameter of Mercury is 3031 mi. What is its circumference? ______________________________

22. What is the circumference of Jupiter if its equatorial radius is 44,000 mi? ______________________________


Copyright © William H. Sadlier, Inc. All rights reserved.

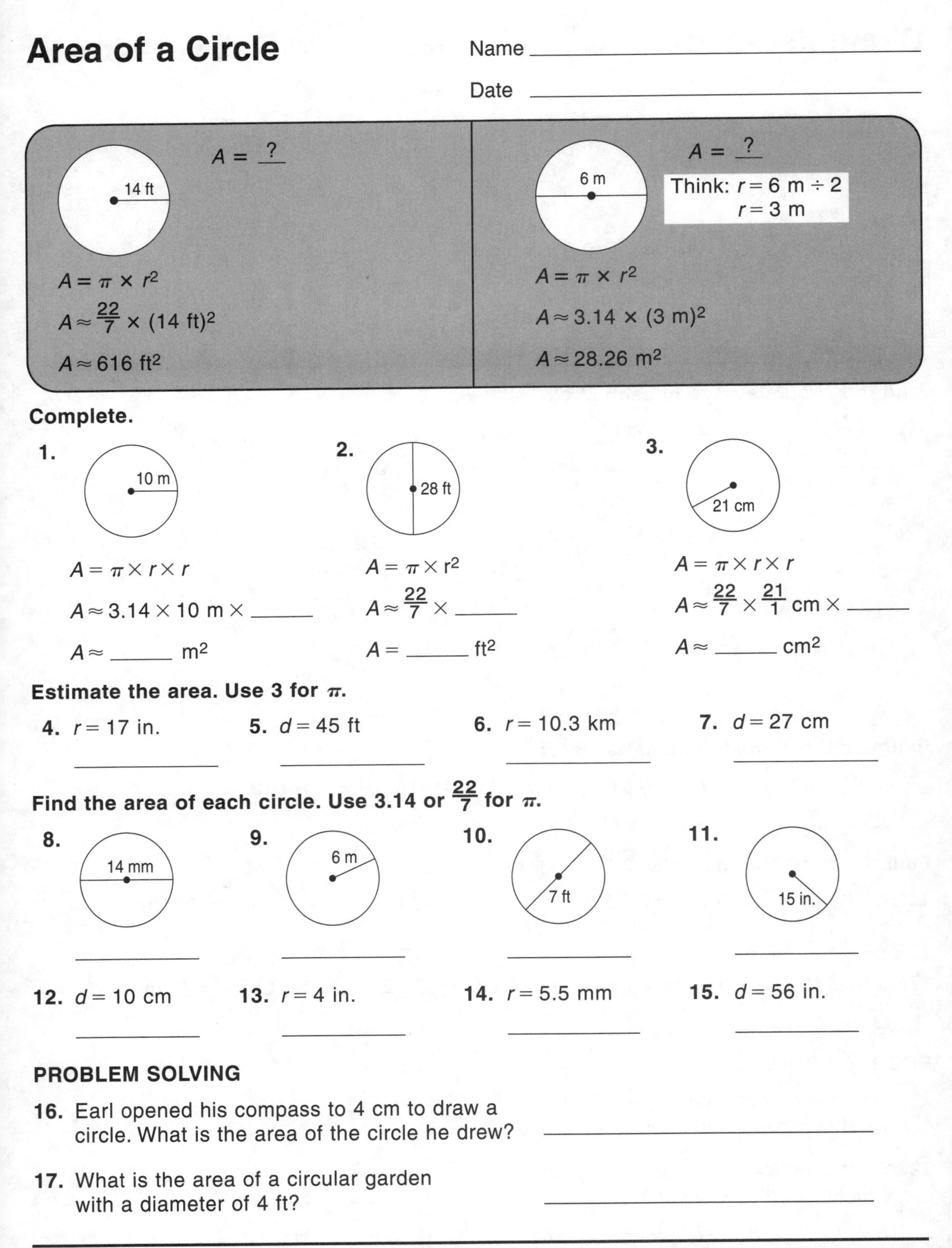

# Area of a Circle

Name ______________________

Date ______________________

$A = \underline{?}$ (circle with radius 14 ft)

$A = \pi \times r^2$

$A \approx \frac{22}{7} \times (14 \text{ ft})^2$

$A \approx 616 \text{ ft}^2$

$A = \underline{?}$ (circle with diameter 6 m)

Think: $r = 6 \text{ m} \div 2$
$r = 3 \text{ m}$

$A = \pi \times r^2$

$A \approx 3.14 \times (3 \text{ m})^2$

$A \approx 28.26 \text{ m}^2$

**Complete.**

**1.** (10 m)

$A = \pi \times r \times r$

$A \approx 3.14 \times 10 \text{ m} \times$ ______

$A \approx$ ______ $\text{m}^2$

**2.** (28 ft)

$A = \pi \times r^2$

$A \approx \frac{22}{7} \times$ ______

$A =$ ______ $\text{ft}^2$

**3.** (21 cm)

$A = \pi \times r \times r$

$A \approx \frac{22}{7} \times \frac{21}{1} \text{ cm} \times$ ______

$A \approx$ ______ $\text{cm}^2$

**Estimate the area. Use 3 for $\pi$.**

**4.** $r = 17$ in. ______

**5.** $d = 45$ ft ______

**6.** $r = 10.3$ km ______

**7.** $d = 27$ cm ______

**Find the area of each circle. Use 3.14 or $\frac{22}{7}$ for $\pi$.**

**8.** (14 mm) ______

**9.** (6 m) ______

**10.** (7 ft) ______

**11.** (15 in.) ______

**12.** $d = 10$ cm ______

**13.** $r = 4$ in. ______

**14.** $r = 5.5$ mm ______

**15.** $d = 56$ in. ______

**PROBLEM SOLVING**

**16.** Earl opened his compass to 4 cm to draw a circle. What is the area of the circle he drew? ______

**17.** What is the area of a circular garden with a diameter of 4 ft? ______


Copyright © William H. Sadlier, Inc. All rights reserved.

# Volume of a Prism

Name ______________________

Date ______________________

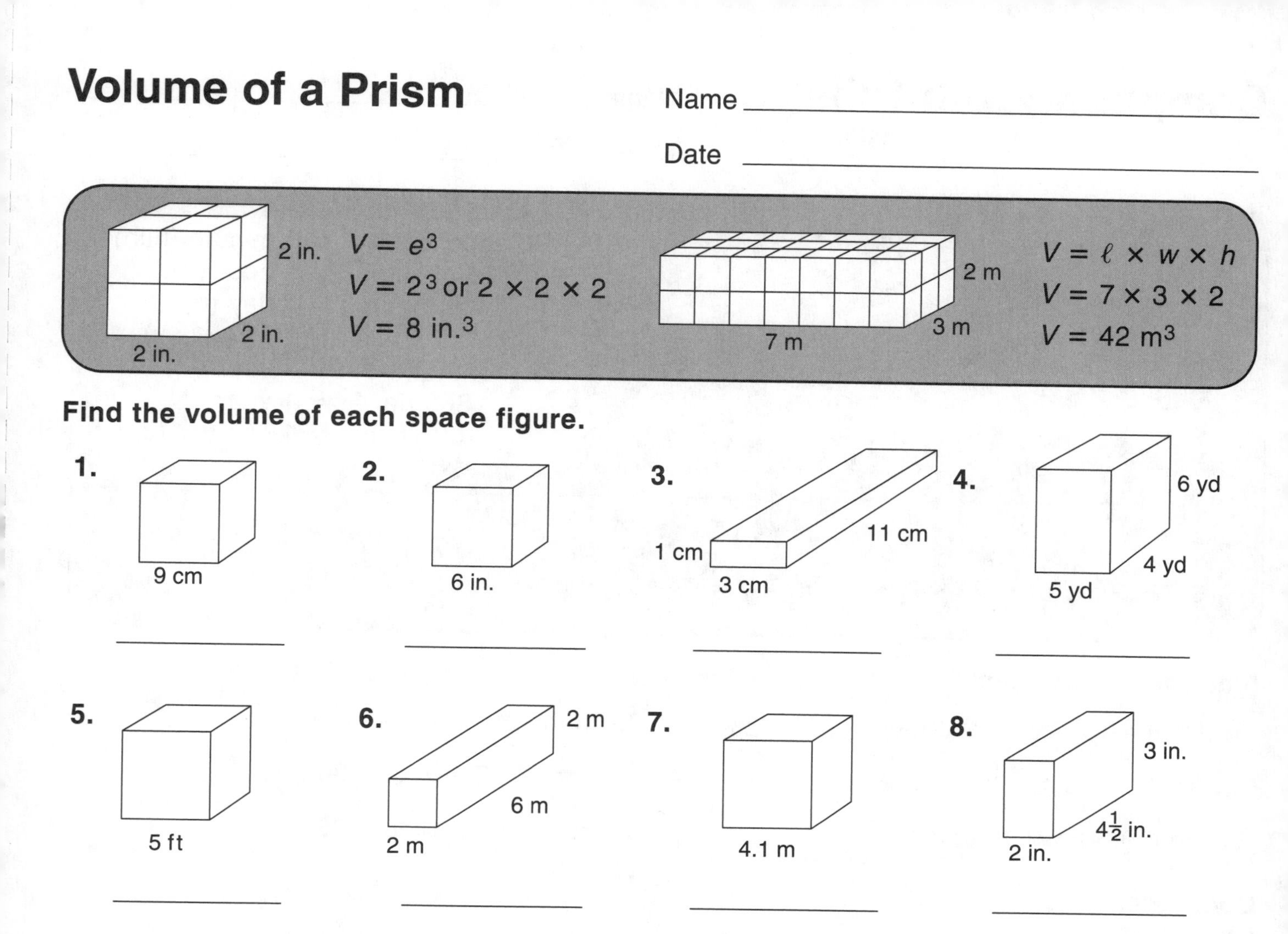


**Find the volume of each space figure.**

1. ________ 2. ________ 3. ________ 4. ________

5. ________ 6. ________ 7. ________ 8. ________

**Complete each table.** Estimate to help you.

| | Rectangular Prism | | | |
|---|---|---|---|---|
| | length | width | height | volume |
| 9. | 4 cm | 1.2 cm | 3.1 cm | |
| 10. | 3 ft | $1\frac{1}{4}$ ft | $2\frac{1}{3}$ ft | |
| 11. | 10 m | 8.4 m | 5 m | |

| | Cube | |
|---|---|---|
| | edge | volume |
| 12. | $7\frac{1}{2}$ ft | |
| 13. | 1.2 cm | |
| 14. | 6.1 mm | |

**PROBLEM SOLVING**

15. Find the volume of a cube-shaped box that measures 40 cm along one edge. ________

16. A box is 10 in. long, 8 in. wide, and 2 in. high. How many cubes that measure 1 in. on each edge will fit inside the box? ________

17. Find the volume of a swimming pool that is 24 ft long, 15 ft wide, and 5 ft deep. ________

 Copyright © William H. Sadlier, Inc. All rights reserved. 

# Computing with Time

Name ______________________

Date ______________________

Find the elapsed time from 2:48 P.M. until 5:05 P.M.

$$\begin{array}{r} 5{:}05 \\ -\ 2{:}48 \\ \hline \end{array} \qquad \begin{array}{r} 4{:}65 \\ \not{5}{:}\not{0}5 \\ -\ 2{:}48 \\ \hline \end{array} \qquad \begin{array}{r} 5\ 15 \\ 4{:}\not{6}\not{5} \\ \not{5}{:}\not{0}5 \\ -\ 2{:}48 \\ \hline 2{:}17 \end{array}$$

The elapsed time is 2 h 17 min.

| | | |
|---|---|---|
| 60 seconds (s) | = | 1 minute (min) |
| 60 minutes | = | 1 hour (h) |
| 24 hours | = | 1 day (d) |
| 7 days | = | 1 week (wk) |
| 12 months (mo) | = | 1 year (y) |
| 365 days | = | 1 year |
| 100 years | = | 1 century (c) |

$$\begin{array}{r} 13\text{ h }45\text{ min} \\ +\ \ 5\text{ h }30\text{ min} \\ \hline 18\text{ h }75\text{ min} \end{array} = 19\text{ h }15$$

$$\begin{array}{r} 4\text{ y }9\text{ mo} \\ \times\quad 5 \\ \hline 20\text{ y }45\text{ mo} \end{array} = 23\text{ y }9\text{ mo}$$

5 wk 1 d ÷ 4
36 d ÷ 4 = 9 d
9 d = 1 wk 2 d

**Find the elapsed time.**

1. from 4:15 A.M. to 9:30 A.M. ____________
2. from 1:45 P.M. to 11:00 P.M. ____________
3. from 11:18 A.M. to 3:20 P.M. ____________
4. from 7:12 A.M. to 1:30 P.M. ____________
5. from 9:30 A.M. to 4:10 P.M. ____________
6. from 8:44 A.M. to 2:35 P.M. ____________

**Complete.**

7. 2 y = ______ mo
8. 120 min = ____ h
9. 21 d = ____ wk
10. 240 s = ______ min
11. 30 mo = ____ y
12. 3 y = _____ d
13. $3\frac{1}{4}$ h = ______ min
14. 425 y = _____ c
15. 192 h = ____ d
16. 128 h = ___ d ___ h
17. 90 d = ___ wk ___ d
18. 525 min = ___ h ___ min

**Compute.**

19. $\begin{array}{r} 8\text{ h }25\text{ min} \\ +\ 4\text{ h }15\text{ min} \\ \hline \end{array}$

20. $\begin{array}{r} 12\text{ h }38\text{ min} \\ -\ \ 8\text{ h }15\text{ min} \\ \hline \end{array}$

21. $\begin{array}{r} 6\text{ h }45\text{ min} \\ +\ 7\text{ h }35\text{ min} \\ \hline \end{array}$

22. $\begin{array}{r} 10\text{ wk }2\text{ d} \\ -\ \ 2\text{ wk }5\text{ d} \\ \hline \end{array}$

23. $\begin{array}{r} 4\text{ d }9\text{ h} \\ \times\quad 2 \\ \hline \end{array}$

24. $\begin{array}{r} 3\text{ d }8\text{ h} \\ \times\quad 7 \\ \hline \end{array}$

25. 18 wk 6 d ÷ 3 ________
26. 9 c 24 y ÷ 4 ________
27. 25 min 4 s ÷ 8 ________

**PROBLEM SOLVING**

28. Ohura left for Dodge City at 8:35 A.M. She arrived in Dodge City at 2:10 P.M. How long was her trip? ________________

 Copyright © William H. Sadlier, Inc. All rights reserved.

# Problem-Solving Strategy: Use Drawings/Formulas

Name ______________________

Date ______________________

A swimming pool is 60 ft long and 30 ft wide.
A 10-foot wide paved walkway surrounds the pool.
What is the perimeter of the outside edge of the walkway?

Use a picture and a formula to solve the problem.

The length is $10 + 60 + 10$ or 80 feet.

The width is $10 + 30 + 10$ or 50 feet.

The formula is $P = 2 \times \ell + 2 \times w$.

The perimeter is $(2 \times 80) + (2 \times 50)$ or 260 feet.

10 ft | 30 ft | 10 ft
10 ft | 60 ft | 10 ft

**PROBLEM SOLVING Do your work on a separate sheet of paper.**

1. Four sticks of margarine are packed in a box that is 12.2 cm long, 6.8 cm wide, and 6.8 cm high. To the nearest hundredth, what is the volume of each stick of margarine?

   ______________________

2. A shed shaped like a rectangular prism is 10 ft long, 5 ft wide, and 7 ft high. If you cover the four outside walls with cedar shingles, how many square feet will be covered?

   ______________________

3. Jon's rectangular garden is 12 ft long and 8 ft wide. There is a stone walk on all four sides of the garden. The stones are 2 feet square. What is the distance around the outer edge of the walk?

   ______________________

4. The diameter of Philip's circular fish pond is 2.4 m. A concrete border around the pond is 0.3 m wide. What is the area of the fish pond? To the nearest hundredth, what is the area of the border around the fish pond?

   ______________________

5. What is the area of the garden described in problem 3? What is the area of the stone walk around the garden?

   ______________________

6. What is the circumference of the border around the fish pond described in Problem 4?

   ______________________

7. Each side of a square park measures 12 yd. Along one side is a triangular playground. Its height is the side of the square, and its base is 6 yd long. What is the combined area of the park and the playground?

   ______________________

8. Sherman pasted 6 square photos on a rectangular sheet of posterboard that is 12 in. long and 8 in. wide. One side of each photo measures 3 in. What area of the posterboard is left showing?

   ______________________


Copyright © William H. Sadlier, Inc. All rights reserved.

# Ratio

Name ______________________

Date ______________________

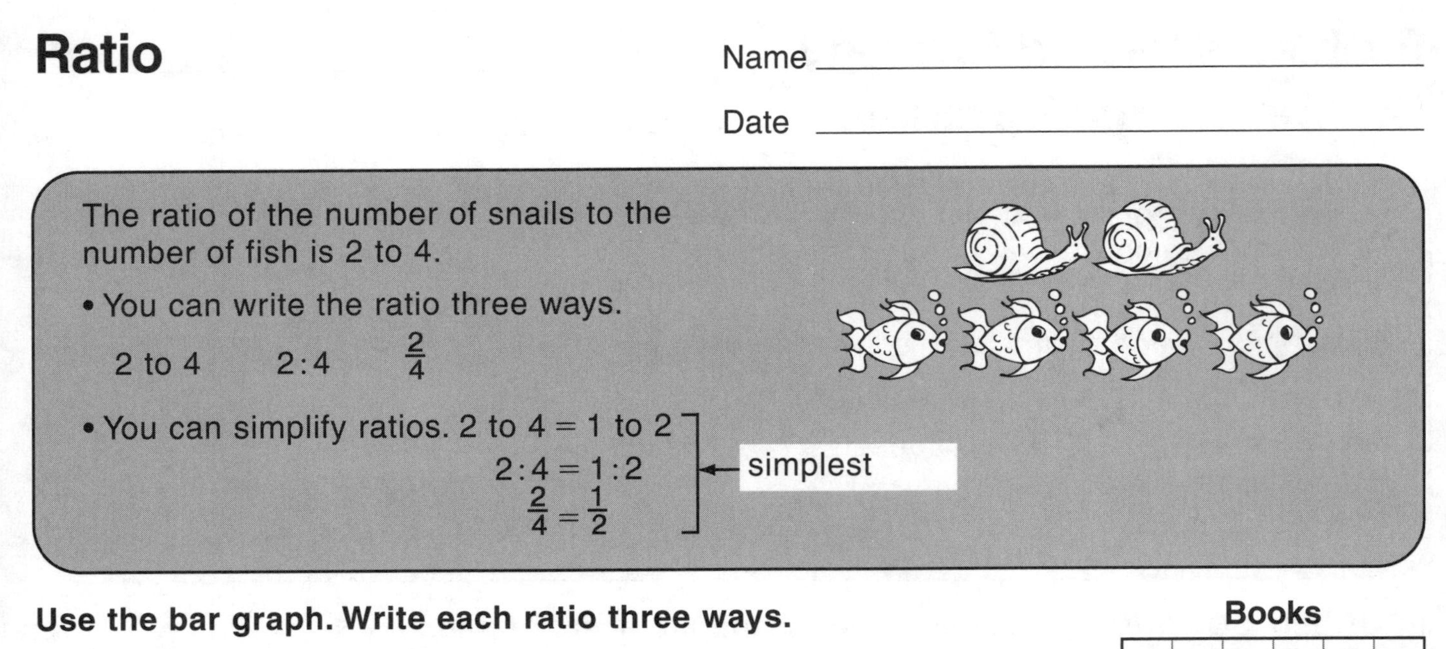

The ratio of the number of snails to the number of fish is 2 to 4.

- You can write the ratio three ways.

  2 to 4     2:4     $\frac{2}{4}$

- You can simplify ratios. 2 to 4 = 1 to 2

  2:4 = 1:2    ← simplest

  $\frac{2}{4} = \frac{1}{2}$

**Use the bar graph. Write each ratio three ways.**

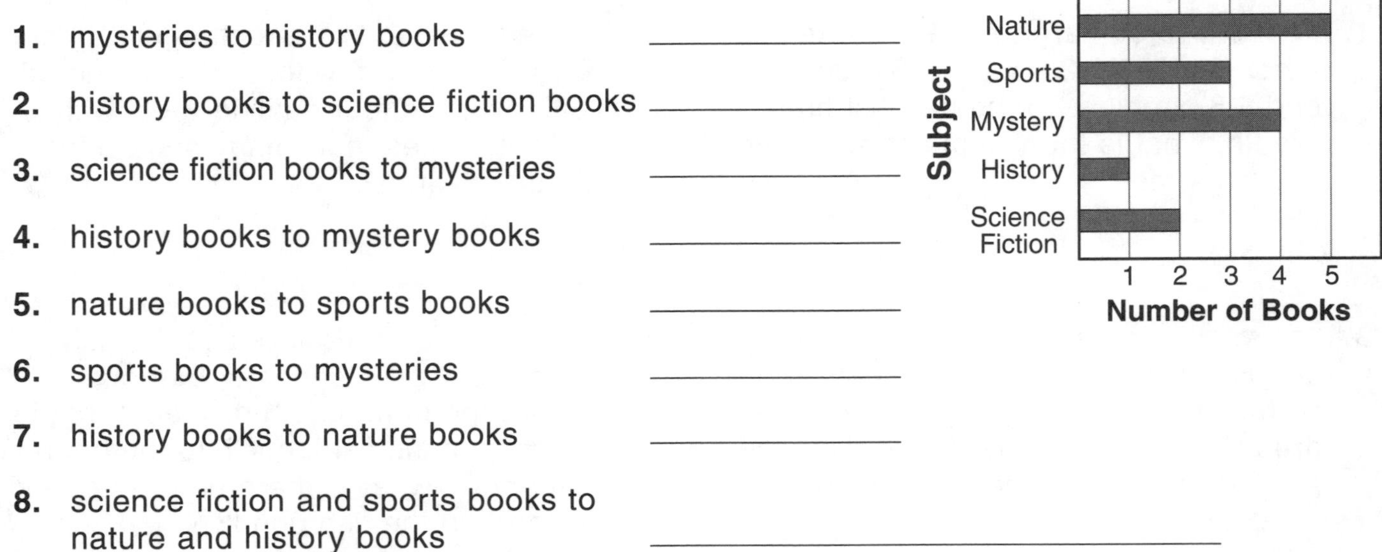


1. mysteries to history books ______________
2. history books to science fiction books ______________
3. science fiction books to mysteries ______________
4. history books to mystery books ______________
5. nature books to sports books ______________
6. sports books to mysteries ______________
7. history books to nature books ______________
8. science fiction and sports books to nature and history books ______________

**Write each ratio in simplest form.**

**9.** 9 to 15 ______ **10.** 4 to 8 ______ **11.** 16 to 20 ______

**12.** 12:3 ______ **13.** $\frac{4}{10}$ ______ **14.** $\frac{21}{78}$ ______

**15.** $\frac{3}{9}$ ______ **16.** 10 to 50 ______ **17.** 8:28 ______

**18.** 6:15 ______ **19.** $\frac{12}{30}$ ______ **20.** 8 to 12 ______

**21.** 1 dime to 1 dollar ______ **22.** 1 pint to 1 quart ______

**23.** 3 inches to 1 foot ______ **24.** 6 ounces to 1 pound ______

**25.** 2 nickels to 6 dimes ______ **26.** 2 dimes to 2 quarters ______

**27.** 2 cups to 2 quarts ______ **28.** 12 inches to 1 yard ______

**29.** 1 quarter to 5 dimes ______ **30.** 3 nickels to 1 dollar ______

 **Use with Lesson 11-1, text pages 376–377.** Copyright © William H. Sadlier, Inc. All rights reserved.

# Equal Ratios

Name ____________________

Date ____________________

Multiply or divide the numerator and the denominator by the same number to find an equal ratio.

| Ratio | | Equal Ratio |
|---|---|---|
| $\frac{6}{8} = \frac{6 \div 2}{8 \div 2}$ | = | $\frac{3}{4}$ |
| $\frac{3}{5} = \frac{3 \times 4}{5 \times 4}$ | = | $\frac{12}{20}$ |

**Write three equal ratios for each.**

1. $\frac{3}{8}$ ________ 2. $\frac{18}{24}$ ________ 3. $\frac{4}{10}$ ________

4. $\frac{8}{16}$ ________ 5. $\frac{15}{9}$ ________ 6. $\frac{1}{2}$ ________

**Circle the letter of the ratio that is equal to the given ratio.**

7. $\frac{2}{3}$ **a.** $\frac{5}{6}$ **b.** 6:9 **c.** 1 to 6 **d.** 6 to 2

8. 4 to 12 **a.** 12:24 **b.** $\frac{1}{4}$ **c.** 5:13 **d.** 1 to 3

9. 10:2 **a.** 5 to 1 **b.** $\frac{5}{2}$ **c.** 1:5 **d.** $\frac{10}{5}$

10. 12 to 8 **a.** $\frac{2}{3}$ **b.** $\frac{3}{2}$ **c.** $\frac{6}{3}$ **d.** 10:6

**Which ratios are equal? Write = or ≠.**

11. $\frac{3}{5}$ ____ $\frac{6}{10}$ 12. $\frac{9}{2}$ ____ $\frac{18}{1}$ 13. $\frac{5}{6}$ ____ $\frac{25}{30}$ 14. $\frac{3}{4}$ ____ $\frac{24}{32}$

15. $\frac{2}{3}$ ____ $\frac{4}{8}$ 16. $\frac{7}{8}$ ____ $\frac{21}{25}$ 17. $\frac{7}{2}$ ____ $\frac{14}{49}$ 18. $\frac{8}{2}$ ____ $\frac{24}{6}$

**Complete to make equal ratios.**

19. $\frac{9}{18} = \frac{\quad}{2}$ 20. $\frac{4}{8} = \frac{8}{\quad}$ 21. $\frac{3}{\quad} = \frac{9}{15}$ 22. $\frac{4}{6} = \frac{\quad}{3}$

23. $\frac{8}{20} = \frac{4}{\quad}$ 24. $\frac{5}{3} = \frac{25}{\quad}$ 25. $\frac{14}{\quad} = \frac{2}{3}$ 26. $\frac{5}{\quad} = \frac{10}{8}$

## PROBLEM SOLVING

27. There are 24 videotapes in 8 identical packages. How many videotapes are in one package? ________

28. There are 12 books in one carton. How many books are in 5 cartons that are the same size? ________

 Copyright © William H. Sadlier, Inc. All rights reserved. 

# Rates

Name ______________________

Date ______________________

A **rate** is a ratio that compares two different quantities.

80 breaths in two minutes $\frac{80 \text{ breaths}}{2 \text{ minute}} = \frac{? \text{ breaths}}{1 \text{ minute}} \longrightarrow \frac{80 \div 2}{2 \div 2} = \frac{40}{1} \longleftarrow$ 40 breaths in 1 minute: unit rate

**Write each rate in simplest form.**

**1.** $\frac{64 \text{ feet}}{2 \text{ seconds}} = \frac{\_\_\_ \text{ feet}}{\text{second}}$ **2.** $\frac{12 \text{ apples}}{4 \text{ children}} = \frac{\_\_\_ \text{ apples}}{\text{child}}$ **3.** $\frac{24 \text{ crayons}}{3 \text{ boxes}} = \frac{\_\_\_ \text{ crayons}}{\text{box}}$

**4.** $\frac{150 \text{ pages}}{2 \text{ hours}} = \frac{\_\_\_ \text{ pages}}{\text{hour}}$ **5.** $\frac{8 \text{ quarts}}{16 \text{ pints}} = \frac{\_\_\_ \text{ quart}}{\text{pints}}$ **6.** $\frac{165 \text{ miles}}{3 \text{ hours}} = \frac{\_\_\_ \text{ miles}}{\text{hour}}$

**Find the unit rate or unit cost. Use equal ratios.**

**7.** 40 meters in 5 seconds ______ **8.** 135 miles in 3 hours ______

**9.** 6 pears for $2.10 ______ **10.** 3 books for $8.85 ______

**11.** 24 pens in 2 boxes ______ **12.** $44 for 8 hours ______

**Use the unit rate or the unit cost to complete.**

**13.** $4 for 1 ticket

______ for 3 tickets

**14.** 30 miles in 1 hour

______ in 5 hours

**15.** 1 pencil for 20¢

3 pencils for ______

**16.** 80 words in 1 minute

______ in 5 minutes

**17.** 1 book for $3.50

4 books for ______

**18.** 32 miles on 1 gallon

______ on 12 gallons

**PROBLEM SOLVING**

**19.** A pilot flew his plane 225 mi in 45 minutes. What was his speed per minute? ______

**20.** If Peter saves $30 in 6 weeks, how much does he save each week? ______

**21.** A 3-lb loaf of bread costs $5.25. At the same rate per pound, how much would a 1-lb loaf cost? ______

**22.** Maria earns $6.50 per hour. How much does she earn working 20 hours? ______

**23.** Mr. Gomez drove 450 mi in 9 hours. What was his average rate of speed per hour? ______

 Copyright © William H. Sadlier, Inc. All rights reserved.

# Proportions

Name ____________________

Date ____________________

A **proportion** is a number sentence stating that two ratios or two rates are equal.

$$\frac{3}{12} \stackrel{?}{=} \frac{5}{20}$$

$$\frac{3}{12} = \frac{1}{4} \text{ and } \frac{5}{20} = \frac{1}{4}$$

$$\frac{3}{12} = \frac{5}{20}$$

$$\frac{2}{5} \stackrel{?}{=} \frac{10}{25}$$

$$2 \times 25 = 5 \times 10$$

$$50 = 50$$

**Write = or ≠. Use equivalent fractions or the cross-products rule.**

1. $\frac{2}{5}$ ______ $\frac{5}{2}$
2. $\frac{4}{7}$ ______ $\frac{8}{14}$
3. $\frac{9}{4}$ ______ $\frac{12}{27}$
4. $\frac{6}{4}$ ______ $\frac{12}{8}$
5. $\frac{6}{10}$ ______ $\frac{3}{5}$
6. $\frac{2}{9}$ ______ $\frac{12}{54}$
7. $\frac{24}{18}$ ______ $\frac{4}{2}$
8. $\frac{16}{6}$ ______ $\frac{3}{8}$
9. $\frac{12}{50}$ ______ $\frac{7}{25}$
10. $\frac{3}{8}$ ______ $\frac{9}{24}$
11. $\frac{1}{4}$ ______ $\frac{25}{100}$
12. $\frac{24}{30}$ ______ $\frac{4}{5}$

**Complete the number sentence to form a proportion.**

13. $\frac{3}{7} = \frac{9}{\_\_}$
14. $\frac{2}{5} = \frac{4}{\_\_}$
15. $\frac{6}{\_\_} = \frac{9}{15}$
16. $\frac{6}{12} = \frac{\_\_}{36}$
17. $\frac{7}{3} = \frac{28}{\_\_}$
18. $\frac{12}{54} = \frac{\_\_}{8}$
19. $\frac{3}{5} = \frac{\_\_}{45}$
20. $\frac{18}{24} = \frac{\_\_}{4}$
21. $\frac{5}{8} = \frac{\_\_}{48}$
22. $\frac{10}{3} = \frac{40}{\_\_}$
23. $\frac{20}{100} = \frac{1}{\_\_}$
24. $\frac{25}{75} = \frac{\_\_}{3}$

**Circle the letters of the two equal ratios.**
**Then write a proportion.**

25. **a.** $\frac{1}{3}$ **b.** $\frac{3}{6}$ **c.** $\frac{5}{15}$ ____________________
26. **a.** $\frac{21}{28}$ **b.** $\frac{3}{4}$ **c.** $\frac{4}{3}$ ____________________
27. **a.** $\frac{2}{10}$ **b.** $\frac{4}{5}$ **c.** $\frac{24}{30}$ ____________________
28. **a.** 6:50 **b.** 3:25 **c.** 5:40 ____________________
29. **a.** 48:12 **b.** 20:5 **c.** 12:4 ____________________
30. **a.** 10:15 **b.** 1:3 **c.** 14:21 ____________________


Copyright © William H. Sadlier, Inc. All rights reserved.

# Solving Proportions

Name ____________________

Date ____________________

**Extremes** **Means** Check:

$\frac{n}{12} \times \frac{5}{20} \longrightarrow$ $n \times 20 = 12 \times 5$ $\frac{3}{12} \stackrel{?}{=} \frac{5}{20} \longrightarrow \frac{1}{4} = \frac{1}{4}$

$n \times 20 = 60$

$n = \frac{60}{20}$

$n = 3$

**Complete.**

1. $\frac{n}{8} = \frac{30}{48} \longrightarrow n \times 48 = 8 \times 30$
   $n \times 48 = 240$
   $n = 240 \div$ ____
   $n =$ ____

2. $\frac{7}{n} = \frac{21}{30} \longrightarrow 7 \times 30 = n \times 21$
   $210 = n \times 21$
   $210 \div$ ____ $= n$
   ____ $= n$

**Find the missing term in each proportion.**

3. $\frac{n}{3} = \frac{10}{15}$ ____
4. $\frac{9}{10} = \frac{n}{40}$ ____
5. $\frac{n}{4} = \frac{9}{6}$ ____
6. $\frac{8}{40} = \frac{n}{20}$ ____
7. $\frac{6}{7} = \frac{n}{21}$ ____
8. $\frac{6}{9} = \frac{12}{n}$ ____
9. $\frac{8}{12} = \frac{n}{24}$ ____
10. $\frac{3}{n} = \frac{24}{16}$ ____
11. $\frac{n}{6} = \frac{35}{42}$ ____
12. $\frac{2}{n} = \frac{5}{10}$ ____
13. $\frac{3}{4} = \frac{n}{48}$ ____
14. $\frac{16}{20} = \frac{48}{n}$ ____

**Find the value of *n*.**

15. $4:n = 16:24$ ____
16. $n:7 = 7:49$ ____
17. $8:3 = n:12$ ____
18. $5:n = 15:21$ ____
19. $2:9 = 16:n$ ____
20. $n:12 = 15:60$ ____
21. $9:27 = n:3$ ____
22. $n:5 = 6:30$ ____
23. $12:5 = 4:n$ ____

**Circle the letters of the two ratios that form a proportion.**

24. a. $\frac{1}{3}$ b. $\frac{1}{6}$ c. $\frac{2}{6}$
25. a. $\frac{3}{8}$ b. $\frac{5}{6}$ c. $\frac{10}{12}$
26. a. $\frac{5}{10}$ b. $\frac{10}{15}$ c. $\frac{1}{2}$
27. a. $\frac{12}{30}$ b. $\frac{2}{5}$ c. $\frac{6}{5}$
28. a. $\frac{7}{8}$ b. $\frac{21}{24}$ c. $\frac{14}{21}$
29. a. $\frac{28}{30}$ b. $\frac{35}{50}$ c. $\frac{7}{10}$
30. a. $\frac{16}{56}$ b. $\frac{3}{14}$ c. $\frac{2}{7}$
31. a. $\frac{4}{16}$ b. $\frac{1}{4}$ c. $\frac{1}{8}$

 Use with Lesson 11-5, text pages 384–385. Copyright © William H. Sadlier, Inc. All rights reserved.

# Writing Proportions

Name ______________________

Date ______________________

| Hourly Wages | |
|---|---|
| Laura | $6.00 |
| Tom | $5.75 |
| Tyrell | $6.50 |
| Teresa | $6.25 |
| Wai Kai | $7.00 |

Write a proportion to show how much Tyrell would earn if he works 20 hours.
Let $w =$ total earnings.

hours → / earnings → $\frac{1}{\$6.50} = \frac{20}{w}$ ← hours / ← earnings

or

earnings → / hours → $\frac{\$6.50}{1} = \frac{w}{20}$ ← earnings / ← hours

or

1 hour → / several hours → $\frac{1}{20} = \frac{\$6.50}{w}$ ← 1 hour / ← several hours

**Use the table above. Write a proportion to find the Total earnings for each number of hours worked.**

1. Laura, 25 hours ______________
2. Waikai, 16 hours ______________
3. Tom, 31 hours ______________
4. Teresa, 10 hours ______________
5. Tyrell, 18 hours ______________
6. Laura, 8 hours ______________

**Write a proportion. Use a variable when necessary.**

7. Oscar gets 2 hits out of every 5 times he comes to bat. How many hits would you expect him to have in 60 at bats? ______________

8. Jason averages 4 earned runs for every 9 innings he pitches How many earned runs would he have if he pitched 45 innings? ______________

9. One out of every 37 pitches is hit into the stands. Out of 148 pitches, how many would be hit into the stands? ______________

10. For every 25 hot dogs that Elena sells to the fans, she sells 2 team pennants. How many hot dogs would she have sold if she sold 16 pennants? ______________


Copyright © William H. Sadlier, Inc. All rights reserved.

# Using Proportions

Name ______________________

Date ______________________

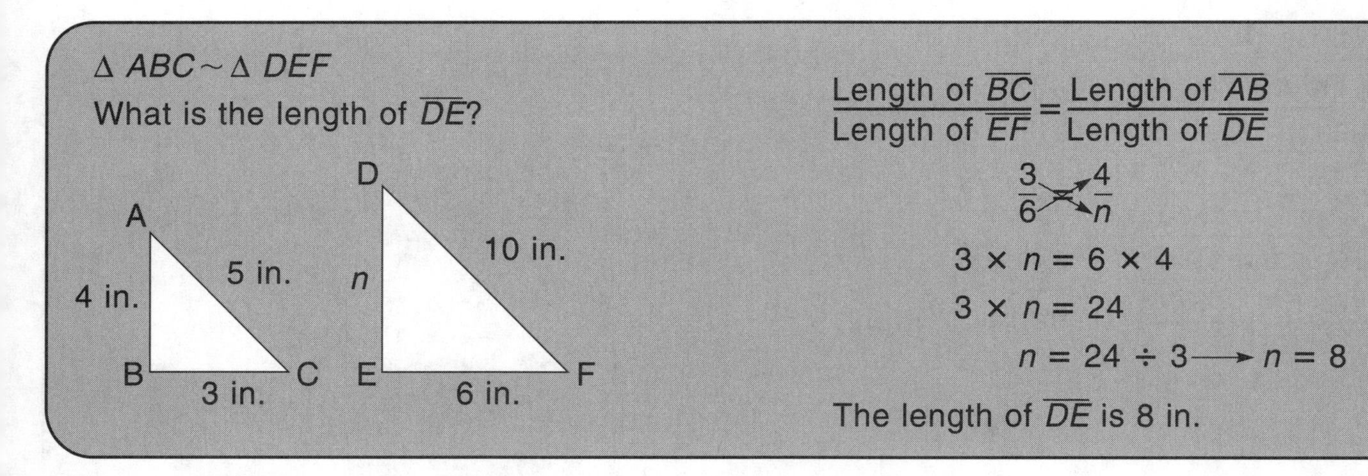


**Write a proportion. Then solve.**

1. In 1 hour Arthur hikes 3 miles. At the same rate, how long will it take him to hike 12 miles? ______________

2. If motor oil sells at 3 quarts for $2.85, what will 5 quarts cost? ______________

3. There are 3 girls for every 2 boys in the Drama Club. If there are 10 boys in the club, how many girls are there? ______________

4. A 10-foot pole casts a 15-foot shadow. At the same time, a tree casts a 24-foot shadow. How tall is the tree? ______________

5. The rectangles are similar. What is the length of $n$?

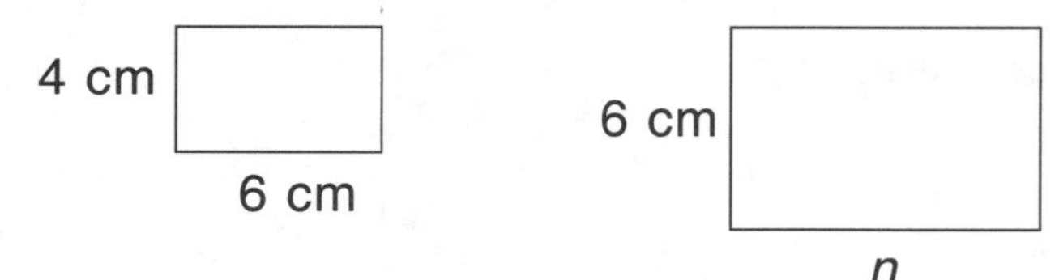


______________

6. A fence post that is 1 meter tall casts a shadow 6 meters long. At the same time, a tree casts a shadow 108 meters. How tall is the tree? ______________

7. Carmen creates a design using 3 squares for every 7 circles. If the design has 12 squares, how many circles are there? ______________

8. Triangle $RST$ is similar to triangle $XYZ$. What is the length of $XY$?

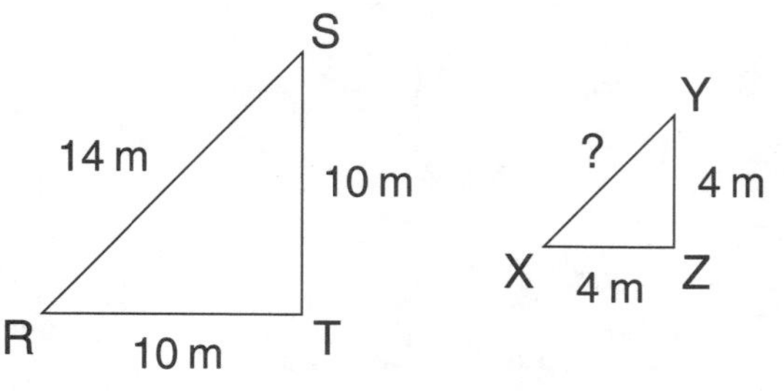


______________

 Copyright © William H. Sadlier, Inc. All rights reserved.

# Scale Drawings and Maps

Name ____________________

Date ____________________

2.8 cm

Scale: 1 cm = 3 km

$\frac{1 \text{ cm}}{3 \text{ km}} = \frac{2.8 \text{ cm}}{? \text{ km}} \longrightarrow \frac{1}{3} = \frac{2.8}{n}$

$1 \times n = 3 \times 2.8$

$n = 8.4$

The actual length is 8.4 km.

**Use a centimeter ruler and the map on the right to complete the table.**

| | To go from: | Scale Distance (cm) | Actual Distance (km) |
|---|---|---|---|
| 1. | North Beach to Lagoon | | |
| 2. | Oak Bluff to Dancing House | | |
| 3. | Town Beach to North Beach | | |
| 4. | Pine Village to Town Beach | | |
| 5. | Lagoon to Surfing | | |
| 6. | Surfing to North Beach | | |
| 7. | Swamp to Lighthouse | | |

North Beach
Surfing
Lagoon
Pine Village
Oak Bluff
Town Beach
Dancing House
Lighthouse
Swamp

**Scale: 1 cm = 10 km**

**Use the scale drawing of the ladybug to answer each question.**

8. What is the scale length of the ladybug's body? ______

9. What is the actual length of the ladybug's body? ______

10. What is the scale width of the ladybug's body? ______

11. What is the actual width of the ladybug's body? ______

**Scale: 1 cm = 5 mm**

**Find the number of actual miles.**
**Scale: 1 in. = 15 mi**

12. 4 in. = ______
13. 10 in. = ______
14. 20 in. = ______
15. 6 in. = ______
16. $5\frac{1}{3}$ in. = ______
17. $2\frac{1}{5}$ in. = ______
18. $3\frac{1}{2}$ in. = ______
19. $1\frac{3}{4}$ in. = ______

**Find the number of actual feet.**
**Scale: 1 in. = 12 ft**

20. 6 in. = ______
21. 5 in. = ______
22. 3 in. = ______
23. 10 in. = ______
24. $2\frac{1}{2}$ in. = ______
25. $13\frac{1}{4}$ in. = ______
26. $10\frac{3}{4}$ in. = ______
27. 25 in. = ______


Copyright © William H. Sadlier, Inc. All rights reserved.

# Percent as Ratio

Name ______________________

Date ______________________

Ratios — Percent

43 to 100 = 43:100 = $\frac{43}{100}$ = 43%

**Write a ratio to show the part of the grid that is shaded. Then write the ratio as a percent.**

**1.** $\frac{\quad}{100}$ = ____

**2.** $\frac{\quad}{100}$ = ____

**3.** $\frac{\quad}{100}$ = ____

**4.** $\frac{\quad}{100}$ = ____

**Write each ratio as a percent.**

**5.** $\frac{25}{100}$ = ____ **6.** $\frac{32}{100}$ = ____ **7.** $\frac{16}{100}$ = ____ **8.** $\frac{73}{100}$ = ____

**9.** $\frac{28}{100}$ = ____ **10.** $\frac{8}{100}$ = ____ **11.** $\frac{82}{100}$ = ____ **12.** $\frac{100}{100}$ = ____

**13.** $\frac{3}{100}$ = ____ **14.** $\frac{99}{100}$ = ____ **15.** 12 to 100 ____ **16.** 17:100 ____

**17.** 61:100 ____ **18.** 98 to 100 ____ **19.** 23 to 100 ____ **20.** 56:100 ____

**21.** 63:100 ____ **22.** 6 to 100 ____ **23.** 99 to 100 ____ **24.** 1 to 100 ____

**Write each percent as a ratio in three ways.**

**25.** 80% ____________ **26.** 75% ____________

**27.** 17% ____________ **28.** 2% ____________

**29.** 100% ____________ **30.** 93% ____________

**31.** 66% ____________ **32.** 1% ____________

**33.** 12% ____________ **34.** 10% ____________

**PROBLEM SOLVING**

**35.** Twenty-seven out of 100 tiles in a design are orange. What percent are orange? ____________

**36.** Of 100 students, 36 are in the debate club. What percent of the students are in the debate club? What percent are not in the debate club? ____________


Copyright © William H. Sadlier, Inc. All rights reserved.

# Relating Percents to Fractions

Name ____________________

Date ____________________

Rename 75% as a fraction.

$75\% = \frac{75}{100} = \frac{75 \div 25}{100 \div 25} = \frac{3}{4}$

Rename $\frac{3}{5}$ as a percent.

$\frac{3}{5} = \frac{3 \times 20}{5 \times 20} = \frac{60}{100} = 60\%$

**Complete.**

**1.** $\frac{7}{10} = \frac{\quad}{100} =$ ______ %

**2.** $\frac{11}{20} = \frac{\quad}{100} =$ ______ %

**3.** $\frac{2}{5} = \frac{\quad}{100} =$ ______ %

**4.** $65\% = \frac{\quad}{100} = \frac{\quad}{20}$

**5.** $24\% = \frac{\quad}{100} = \frac{6}{\quad}$

**6.** $25\% = \frac{\quad}{100} = \frac{1}{\quad}$

**Write as a percent.**

**7.** $\frac{19}{100}$ ______

**8.** $\frac{37}{100}$ ______

**9.** $\frac{3}{10}$ ______

**10.** $\frac{1}{10}$ ______

**11.** $\frac{2}{5}$ ______

**12.** $\frac{1}{2}$ ______

**13.** $\frac{9}{25}$ ______

**14.** $\frac{18}{20}$ ______

**15.** $\frac{38}{50}$ ______

**16.** $\frac{4}{5}$ ______

**17.** $\frac{25}{40}$ ______

**18.** $\frac{15}{75}$ ______

**Write as a fraction in simplest form.**

**19.** 80% ______

**20.** 20% ______

**21.** 11% ______

**22.** 32% ______

**23.** 2% ______

**24.** 88% ______

**25.** 90% ______

**26.** 74% ______

**27.** 43% ______

**28.** 66% ______

**29.** 7% ______

**30.** 15% ______

**Estimate what percent of each figure is shaded.**

**31.** ______________

**32.** ______________

**33.** ______________

**34.** ______________


Copyright © William H. Sadlier, Inc. All rights reserved.

# Relating Percents to Decimals

Name ______________________________

Date ______________________________

Rename 6% as a decimal.

6% = 06. = 0.06

Move the decimal point 2 places to the *left*.

Rename 0.12 as a percent.

0.12 = 0.12 = 12%

Move the decimal point 2 places to the *right*.

**Write as a decimal.**

**1.** 26% ______ **2.** 54% ______ **3.** 14% ______ **4.** 50% ______

**5.** 95% ______ **6.** 73% ______ **7.** 42% ______ **8.** 37% ______

**9.** 7% ______ **10.** 2% ______ **11.** 5% ______ **12.** 9% ______

**Write as a percent.**

**13.** 0.35 ______ **14.** 0.86 ______ **15.** 0.51 ______ **16.** 0.23 ______

**17.** 0.1 ______ **18.** 0.6 ______ **19.** 0.4 ______ **20.** 0.8 ______

**21.** 0.05 ______ **22.** 0.03 ______ **23.** 0.01 ______ **24.** 0.06 ______

**Complete each table. Write each fraction in simplest form.**

| | Percent | Decimal | Fraction |
|---|---|---|---|
| **25.** | 30% | | |
| **26.** | 75% | | |
| **27.** | | 0.2 | |
| **28.** | | | $\frac{11}{20}$ |
| **29.** | | | $\frac{2}{25}$ |

| | Percent | Decimal | Fraction |
|---|---|---|---|
| **30.** | | 0.05 | |
| **31.** | 38% | | |
| **32.** | | 0.65 | |
| **33.** | 40% | | |
| **34.** | | | $\frac{19}{20}$ |

**PROBLEM SOLVING**

**35.** In a survey of 100 people, 7 out of every 25 said that Abraham Lincoln was the greatest U.S. President. What percent of the people does this represent? ______________________


Copyright © William H. Sadlier, Inc. All rights reserved.

# Decimals, Fractions, and Percents

Name ____________________

Date ____________________

Rename 37.5% as a decimal.

37.5% = 0.37.5 = 0.375

Rename $\frac{7}{10}$ as a percent.

$\frac{7}{10} \rightarrow 10\overline{)7.00}$ = 0.70 $\rightarrow$ 70%

or

$\frac{7}{10} = \frac{7 \times 10}{10 \times 10} = \frac{70}{100} = 70\%$

**Write as a decimal.**

| | | | |
|---|---|---|---|
| **1.** 16.3% ______ | **2.** 41.2% ______ | **3.** 56% ______ | **4.** 28.9% ______ |
| **5.** 3% ______ | **6.** 12.54% ______ | **7.** 1.1% ______ | **8.** 70.5% ______ |
| **9.** 6% ______ | **10.** 73.45% ______ | **11.** 4.01% ______ | **12.** 3.5% ______ |

**Write as a percent.**

| | | | |
|---|---|---|---|
| **13.** $\frac{6}{10}$ ______ | **14.** $\frac{4}{5}$ ______ | **15.** $\frac{3}{8}$ ______ | **16.** $\frac{5}{16}$ ______ |
| **17.** $\frac{3}{16}$ ______ | **18.** $\frac{18}{40}$ ______ | **19.** $\frac{9}{20}$ ______ | **20.** $\frac{13}{20}$ ______ |
| **21.** $\frac{45}{80}$ ______ | **22.** $\frac{15}{24}$ ______ | **23.** $\frac{13}{25}$ ______ | **24.** $\frac{30}{48}$ ______ |

**Write as a fractional percent.**

| | | | |
|---|---|---|---|
| **25.** $\frac{7}{9}$ ______ | **26.** $\frac{2}{3}$ ______ | **27.** $\frac{7}{8}$ ______ | **28.** $\frac{5}{12}$ ______ |
| **29.** $\frac{4}{9}$ ______ | **30.** $\frac{4}{7}$ ______ | **31.** $\frac{7}{12}$ ______ | **32.** $\frac{5}{8}$ ______ |
| **33.** $\frac{5}{7}$ ______ | **34.** $\frac{2}{15}$ ______ | **35.** $\frac{1}{45}$ ______ | **36.** $\frac{7}{15}$ ______ |

**PROBLEM SOLVING**

**37.** The math team correctly answered $\frac{9}{10}$ of its questions. What percent did the team answer correctly? ____________

**38.** Marion read $\frac{1}{3}$ of the book. What percent of the book did she read? ____________

**39.** Harvey completed $\frac{5}{8}$ of his homework. What percent of his homework did he complete? ____________

 Copyright © William H. Sadlier, Inc. All rights reserved. 

# Percents Greater Than 100%

Name ______________________

Date ______________________

| Rename 125% and 310% as decimals. | Rename 125% and 310% as mixed numbers. |
|---|---|
| 125% = 1.25. = 1.25 | $125\% = \frac{125}{100} = 1\frac{25}{100} = 1\frac{1}{4}$ |
| 310% = 3.10. = 3.1 | $310\% = \frac{310}{100} = 3\frac{10}{100} = 3\frac{1}{10}$ |

**Write each as a decimal.**

**1.** 135% ______ **2.** 340% ______ **3.** 475% ______ **4.** 301% ______

**5.** 202% ______ **6.** 751% ______ **7.** 915% ______ **8.** 427% ______

**Write each as a mixed number in simplest form.**

**9.** 107% ______ **10.** 410% ______ **11.** 254% ______ **12.** 850% ______

**13.** 375% ______ **14.** 187% ______ **15.** 550% ______ **16.** 322% ______

**Complete each table. Write each mixed number in simplest form.**

| | Percent | Decimal | Fraction |
|---|---|---|---|
| **17.** | 425% | | |
| **18.** | 112% | | |
| **19.** | 250% | | |
| **20.** | 300% | | |
| **21.** | 440% | | |

| | Percent | Decimal | Fraction |
|---|---|---|---|
| **22.** | 108% | | |
| **23.** | | 1.1 | |
| **24.** | | | $4\frac{1}{25}$ |
| **25.** | 932% | | |
| **26.** | 150% | | |

**PROBLEM SOLVING**

**27.** The number 18 is $1\frac{1}{2}$ times 12. Write $1\frac{1}{2}$ as a percent? ______________

**28.** The cost of living is 230% of what it was 10 years ago. What mixed number is this? ______________

**29.** The rainfall this month is 180% of last month's rainfall. What decimal is this? ______________

**30.** This year the cost of a pair of tennis shoes is 35% higher than it was last year. What percent of last year's price is this year's price? ______________

 **Use with Lesson 11-13, text pages 400–401.** Copyright © William H. Sadlier, Inc. All rights reserved.

# Problem-Solving Strategy: Combining Strategies

Name ______________________

Date ______________________

The Johnsons built a rectangular brick patio behind their new home. It is 3 times as long as it is wide. To build the patio, the Johnsons needed enough bricks to cover 675 square feet. What are the dimensions of the patio?

*Use the formula* for the area of a rectangle:
$A = \ell \times w$

The problem tells you that the length is 3 times the width.

*Make a table* and *guess and test* to solve the problem.

The patio is 15 feet wide and 45 feet long.

| | **Guess 1** | **Guess 2** | **Guess 3** | **Guess 4** |
|---|---|---|---|---|
| **Width** | 10 ft | 20 ft | 12 ft | 15 ft |
| **Length** | 30 ft | 60 ft | 36 ft | 45 ft |
| **Area** | 300 ft$^2$ | 1200 ft$^2$ | 432 ft$^2$ | 675 ft$^2$ |

**PROBLEM SOLVING. Do your work on a separate sheet of paper.**

**1.** Liane collected seashells. She gave $\frac{1}{4}$ of the shells to her cousin. The next day she gave 5 shells to her aunt. Then she had 55 shells left. How many shells did Liane start with?

______________________

**2.** Gina worked after school part-time. One day she addressed 50 letters and postcards. Gina noticed that she addressed 1 postcard for every 4 letters. How many postcards did Gina address?

______________________

**3.** Hamid worked 2 hours each day from Monday through Friday. He worked 6 hours on Saturday and did not work on Sunday. At the end of the week, he was paid $76.80. How much per hour did Hamid earn?

______________________

**4.** Elena is making a rectangular dog pen. The dimensions are 20 ft by 16 ft. There is a fence post at each corner and every 4 ft in between. If posts cost $7.95 each, how much will they cost in all?

______________________

**5.** The 15 members of Herb's scout troop are on a weekend camp-out. Herb takes 3 dozen eggs from the ice chest and cooks 2 eggs for each member of the troop. If he returns the unused eggs to the ice chest, how many eggs does he return?

______________________

**6.** There are 56 members of the Golden Age Travel Troup. One-fourth of them are 75 years or older. Fifty percent are between 65 and 75 years old. If $\frac{1}{2}$ of the remaining members are women, how many men in the troup are age 65 or younger?

______________________


Copyright © William H. Sadlier, Inc. All rights reserved.

# Mental Math: Percent

Name ______________________

Date ______________________

Use common fractions to find a percent of a number mentally.

| | | |
|---|---|---|
| 60% of 35 = $\underline{?}$ | 30% of 90 = $\underline{?}$ | $87\frac{1}{2}$% of 56 = $\underline{?}$ |
| Think: 60% = $\frac{3}{5}$ | Think: 30% = $\frac{3}{10}$ | Think: $87\frac{1}{2}$% = $\frac{7}{8}$ |
| $\frac{3}{5}$ of 35 = 21 | $\frac{3}{10}$ of 90 = 27 | $\frac{7}{8}$ of 56 = 49 |
| So 60% of 35 = 21. | So 30% of 90 = 27. | So $87\frac{1}{2}$% of 56 = 49. |

**Complete. Look for patterns.**

1. $\frac{1}{10}$ of 40 is 4, so 10% of 40 is ____.
2. $\frac{2}{10}$ of 40 is 8, so 20% of 40 is ____.
3. $\frac{3}{10}$ of 40 is 12, so 30% of 40 is ____.
4. $\frac{4}{10}$ of 40 is ____, so 40% of 40 is ____.
5. $\frac{1}{8}$ of 64 is 8, so $12\frac{1}{2}$% of 64 is ____.
6. $\frac{1}{4}$ of 64 is 16, so 25% of 64 is ____.
7. $\frac{3}{8}$ of 64 is 24, so $37\frac{1}{2}$% of 64 is ____.
8. $\frac{1}{2}$ of 64 is ____, so 50% of 64 is ____.
9. $\frac{5}{8}$ of 64 is ____, so $62\frac{1}{2}$% of 64 is ____.
10. $\frac{3}{4}$ of 64 is ____, so 75% of 64 is ____.

**Find the percent of the number. Compute mentally.**

11. 20% of 10 ______
12. 50% of 22 ______
13. 30% of 40 ______
14. 60% of 70 ______
15. 25% of 16 ______
16. 75% of 24 ______
17. 40% of 55 ______
18. 10% of 80 ______
19. $12\frac{1}{2}$% of 64 ______
20. $33\frac{1}{3}$% of 21 ______
21. $16\frac{2}{3}$% of 42 ______
22. 80% of 40 ______
23. $37\frac{1}{2}$% of 32 ______
24. $83\frac{1}{3}$% of 36 ______
25. $87\frac{1}{2}$% of 80 ______
26. $83\frac{1}{3}$% of 60 ______
27. $62\frac{1}{2}$% of 88 ______
28. 75% of 56 ______

**PROBLEM SOLVING**

29. Four fifths of the 30 students in Ms. Asai's class watched a special on television last night. What percent of the students watched the special? How many students is this?

______________________


Copyright © William H. Sadlier, Inc. All rights reserved.

# Percent Sense

Name ____________________

Date ____________________

Which is greater:

25% of 20 or 50% of 20?

50% of 20 > 25% of 20

Is $\frac{16}{40}$ less than 25%?

$25\% = \frac{1}{4}$

$\frac{1}{4}$ of 40 = 10

10 is less than 16

So $\frac{16}{40}$ is *not* less than 25%.

**Compare. Write < or >.**

| | | | | | | | |
|---|---|---|---|---|---|---|---|
| **1.** | 25% of 44 | ______ | 25% of 64 | **2.** | 30% of 50 | ______ | 30% of 20 |
| **3.** | 4% of 12 | ______ | 4% of 15 | **4.** | 5% of 20 | ______ | 50% of 20 |
| **5.** | 18% of 45 | ______ | 30% of 45 | **6.** | 85% of 64 | ______ | 20% of 6 |
| **7.** | $12\frac{1}{2}\%$ of 80 | ______ | 19% of 20 | **8.** | $83\frac{1}{3}\%$ of 48 | ______ | $66\frac{2}{3}\%$ of 45 |
| **9.** | $\frac{1}{8}$ of 24 | ______ | $37\frac{1}{2}\%$ of 24 | **10.** | $\frac{5}{6}$ of 18 | ______ | 75% of 18 |
| **11.** | $16\frac{2}{3}\%$ of 42 | ______ | $\frac{1}{3}$ of 42 | **12.** | $\frac{1}{4}$ of 24 | ______ | 20% of 24 |

**Read the situation. Then circle the letter of the correct statement.**

**13.** Lola has one hour to finish her homework. She studies history for 25 minutes.

**a.** Lola studies history for $33\frac{1}{3}\%$ of the hour.

**b.** Lola studies history for less than $33\frac{1}{3}\%$ of the hour.

**c.** Lola studies history for more than $33\frac{1}{3}\%$ of the hour.

**14.** There are 110 entries in the city-wide art competition. Twenty-five of the entries are acrylic paintings.

**a.** Fewer than 25% of the entries are acrylic paintings.

**b.** Fewer than 15% of the entries are acrylic paintings.

**c.** More than 30% of the entries are acrylic paintings.

**15.** Forty out of 60 students received a passing grade on a math test.

**a.** Fewer than 30% of the students did *not* pass the test.

**b.** More than 80% of the students passed the test.

**c.** Exactly $66\frac{2}{3}\%$ of the students passed the test.


Copyright © William H. Sadlier, Inc. All rights reserved.

# Finding a Percent of a Number

Name ______________________

Date ______________________

Find: 27% of \$320

First estimate. 27% of \$320 ⟶ 25% of \$320

$\frac{1}{4}$ of \$320 = \$80

**Use a decimal.**

27% of \$320 = $p$

$0.27 \times \$320 = p$

$\$86.40 = p$

**Use a fraction.**

27% of \$320 = $p$

$\frac{27}{\cancel{100}_5} \times \frac{\$\cancel{320}^{16}}{1} = p$

$\$86.40 = p$

**Use a proportion.**

$\frac{n}{\$320} = \frac{27}{100}$

$n \times 100 = \$320 \times 27$

$n = \$8640 \div 100$

$n = \$86.40$

**Use a decimal to find the percent of the number. Estimate first.**

**1.** 50% of 42 = ________ **2.** 75% of 52 = ________ **3.** 40% of 25 = ________

**4.** 5% of \$14 = ________ **5.** 3% of \$420 = ________ **6.** 8% of \$112 = ________

**7.** 30% of 960 = ________ **8.** 90% of 140 = ________ **9.** 25% of 200 = ________

**10.** 60% of 120 = ________ **11.** 20% of 300 = ________ **12.** 30% of 90 = ________

**Use a fraction to find the percent of the number. Estimate first.**

**13.** 40% of 25 = ________ **14.** 78% of 400 = ________ **15.** 45% of 60 = ________

**16.** 50% of 48 = ________ **17.** 75% of 168 = ________ **18.** 80% of 90 = ________

**19.** 34% of 50 = ________ **20.** 35% of 80 = ________ **21.** 82% of 400 = ________

**22.** 25% of \$60 = ________ **23.** 14% of \$250 = ________ **24.** 70% of \$420 = ________

**Use a proportion to find the percent of the number.**

**25.** 30% of 60 = ________ **26.** 80% of 30 = ________ **27.** 15% of 120 = ________

**28.** 25% of 36 = ________ **29.** 65% of 40 = ________ **30.** 45% of 20 = ________

**31.** 12% of 550 = ________ **32.** 76% of 250 = ________ **33.** 55% of 80 = ________

**34.** 24% of 125 = ________ **35.** 22% of 450 = ________ **36.** 86% of 250 = ________

**PROBLEM SOLVING**

**37.** The distance between two cities is 150 miles. What is 60% of this distance? ______________

**38.** In a basketball game, 37.5% of the 40 foul shots were missed. How many were missed? ______________


Copyright © William H. Sadlier, Inc. All rights reserved.

# Missing Percent

Name ____________________

Date ____________________

What percent of 80 is 32?

**Method 1**

$\frac{32}{80} = r$

$80\overline{)32.00}$ = 0.40

$40\% = r$

So 32 is 40% of 80.

**Method 2**

$\frac{32}{80} = \frac{n}{100}$

$32 \times 100 = 80 \times n$

$n = 3200 \div 80 = 40$

$\frac{n}{100} = \frac{40}{100} = 40\%$

**Find the percent.**

1. What percent of 90 is 27? ______
2. What percent of 240 is 80? ______
3. What percent of 100 is 13? ______
4. What percent of 50 is 50? ______
5. 3.2 is what percent of 80? ______
6. What percent of 28 is 7? ______
7. What percent of 27 is 18? ______
8. What percent of 111 is 444? ______
9. What percent of 150 is 50? ______
10. What percent of 150 is 30? ______
11. What percent of 90 is 18? ______
12. What percent of 120 is 48? ______
13. 2.2 is what percent of 40? ______
14. What percent of 25 is 12? ______
15. What percent of 5 is 10? ______
16. 2.4 is what percent of 25? ______
17. 180 is what percent of 60? ______
18. 16 is what percent of 64? ______

**PROBLEM SOLVING**

19. First Street School won 12 out of 30 awards in the team competition. What percent of the awards did the school win? ______

20. From a group of 280 children, 14 made the swim team. What percent of the children is this? ______

21. Mari had \$2. She spent 50¢ to buy fruit for lunch. What percent of her money did she spend on the fruit? ______

22. Sometimes the doors of Marguerite's train do not open. Of 24 doors on Marguerite's train, 3 doors did not open. What percent of the doors did not open? ______


Copyright © William H. Sadlier, Inc. All rights reserved.

# Using Percent to Solve Problems

Name ____________________

Date ____________________

| 82% of \$41 = $n$ | \$12 out of \$80 = $n$ |
|---|---|
| $0.82 \times \$41 = n$ | $\frac{\$12}{\$80} = n$ |
| \$33.62 = $n$ | $\$80\overline{)\$12.00}$ quotient 0.15     15% = $n$ |

**PROBLEM SOLVING**

1. What is 12% of \$65? ________
2. What is 15% of 120? ________
3. 12 is what percent of 60? ________
4. 48 is what percent of 96? ________
5. What is 35% of 120? ________
6. \$200 is what percent of \$50? ________
7. Twenty-five students began a 3-mile run. Fifteen of the students completed the run. What percent of the students finished the run? ________
8. In a spelling test of 50 words, Joan spelled 43 words correctly. What percent of the words did Joan spell correctly? ________
9. Mr. Butler sells apples. Of the 400 bushels of apples he sells weekly, 85% are red delicious. How many bushels of red delicious does Mr. Butler sell weekly? ________
10. On Saturday, $87\frac{1}{2}\%$ of the students attended the championship game. If there are 728 students in all, how many attended the game? ________
11. Frank had 55 marbles. After a few games, he had 120% of his original number. How many marbles does he have now? ________
12. Jerry earned \$6.00. He spent \$1.50 on a book. What percent of his earnings did he spend? ________
13. The goal for the book drive for the school library was 120 books. The book drive brought in 175% of the goal. How many books were received in the drive? ________
14. Jonathan had \$180.00 in the bank. He spent $62\frac{1}{2}\%$ of it on a bicycle. How much did the bicycle cost? ________
15. Cole has 40 paperback books. Ten of the books are mysteries. What percent of the books are mysteries? ________

 **Use with Lesson 12-5, text pages 420–421.** Copyright © William H. Sadlier, Inc. All rights reserved.

# Finding Discount and Sale Price

Name ____________________

Date ____________________

A book that costs $26.00 is being sold at a 30% rate of discount.

**Discount = Rate of Discount × List Price**

$D = 30\% \times \$26.00$
$D = 0.30 \times \$26.00$
$D = \$7.80$

The discount on the book is $7.80.

**Sale Price = List Price − Discount**

$SP = \$26.00 - \$7.80$
$SP = \$18.20$

The sale price of the book is $18.20.

**Complete the table.**

| | Item | List Price | Rate of Discount | Discount | Sale Price |
|---|---|---|---|---|---|
| **1.** | Bike | $80 | 25% | | |
| **2.** | Chair | $14 | 20% | | |
| **3.** | Radio | $33 | 15% | | |
| **4.** | Jacket | $120 | 12% | | |
| **5.** | Car | $9000 | 8% | | |

**Find the discount and the sale price.**

**6.** List Price: $24
Rate of Discount: 10%

Discount = ________

Sale Price = ________

**7.** List Price: $75
Rate of Discount: 20%

Discount = ________

Sale Price = ________

**8.** List Price: $56
Rate of Discount: 8%

Discount = ________

Sale Price = ________

**9.** List Price: $520
Rate of Discount: 5%

Discount = ________

Sale Price = ________

**PROBLEM SOLVING**

**10.** Ronald bought a notebook. The regular price was $4.50. It was discounted 20%. How much did Ronald pay? ________

**11.** The price of a table is $478. It is on sale at 40% off. What is the discount? What is the sale price? ________

**12.** The rate of discount on a motorcycle is 10%. If the list price is $6299, what is the sale price? ________


Copyright © William H. Sadlier, Inc. All rights reserved.

# Finding Sales Tax and Total Cost

Name ____________________

Date ____________________

A shirt costs \$38.95 plus 6% sales tax. What is the total cost?

| Sales Tax | = | Rate of Sales Tax | × | Marked Price |
|---|---|---|---|---|

$T = 6\% \times \$38.95$

$T = 0.06 \times \$38.95$

$T = 2.337$ or \$2.34.

The sales tax is \$2.34.

| Total Cost | = | Marked Price | + | Sales Tax |
|---|---|---|---|---|

$TC = \$38.95 + \$2.34$

$TC = \$41.29$

The total cost of the shirt is \$41.29.

**Complete the table.**

| | Item | Marked Price | Rate of Sales Tax | Sales Tax | Total Cost |
|---|---|---|---|---|---|
| **1.** | Softball | \$7.98 | 4% | | |
| **2.** | Glove | \$45.25 | 6% | | |
| **3.** | Wooden Bat | \$23.50 | 2% | | |
| **4.** | Thermos | \$12.00 | 5% | | |
| **5.** | Jacket | \$59.95 | 7% | | |

**Find the sales tax and the total price.**

**6.** Price: \$85.50
Rate of Sales Tax: 4%

Sales Tax = ________

Sale Price = ________

**7.** Price: \$300
Rate of Sales Tax: 7%

Sales Tax = ________

Sale Price = ________

**8.** Price: \$129.95
Rate of Sales Tax: 3%

Sales Tax = ________

Sale Price = ________

**9.** Price: \$12,999
Rate of Sales Tax: 5%

Sales Tax = ________

Sale Price = ________

**PROBLEM SOLVING**

**10.** A video game costs \$82.25 plus 6% sales tax. Find the sales tax. ____________

**11.** A video cassette recorder costs \$143.75 plus 8% sales tax. Find the sales tax and the total cost. ____________

**12.** A rug costs \$296 plus $6\frac{1}{2}$% sales tax. What is the total cost for the rug? ____________

 Copyright © William H. Sadlier, Inc. All rights reserved.

# Better Buy

Name ____________________

Date ____________________

Pens cost 4 for $4.41 or 6 for $6.03. Which is the better buy?

$\frac{4}{\$4.41} = \frac{1}{n}$

$4 \times n = \$4.41 \times 1$

$n = \$4.41 \div 4$

$n = \$1.1025$

$n = \$1.10$

$\frac{6}{\$6.03} = \frac{1}{m}$

$6 \times m = \$6.03 \times 1$

$m = \$6.03 \div 6$

$m = \$1.005$

$m = \$1.01$

$\$1.01 < \$1.10$

So 6 pens for $6.03 is a better buy.

**Which is the better buy?**

1. 3 apples for $.75
   9 apples for $2.10
   ____________________
2. 4 cans of juice for $3.00
   A 6-pack for $4.45
   ____________________
3. 1 pair of socks for $1.55
   3 pairs for $4.69
   ____________________
4. 1 dozen roses for $15.25
   $1.25 per rose
   ____________________
5. 8-oz can for 48¢
   6-oz can for 42¢
   ____________________
6. 2-lb box for $1.84
   5-lb box for $4.50
   ____________________

**Estimate to decide which is the better buy.**

7. 1 pair of socks for $2.50
   3 pairs for $9.65
   ____________________
8. 1 can of juice for $.80
   6 cans of juice for $4.75
   ____________________
9. 1 dozen rolls for $3.58
   Rolls: 35¢ each
   ____________________
10. 1 orange for $.35
    Bag of 6 oranges for $2.50
    ____________________
11. 1 dozen pencils for $3.56
    30¢ for 1 pencil
    ____________________
12. 12-oz box of cereal for $2.40
    15-oz box of cereal for $2.99
    ____________________

**PROBLEM SOLVING. Tell which is the better buy for each.**

13. A pair of jeans in Fine Fitters costs $19.95. The same jeans sell in Custom Clothes at 2 pairs for $38.95. ____________________
14. Soap sells for $2.15 a bar, or one box of 5 bars for $10.95. ____________________
15. An 8-oz can of mixed fruit costs $.83. A 14-oz can costs $1.38. ____________________
16. A package of 4 glasses costs $2.98. A package of 6 of the same glasses costs $4.68. ____________________


Copyright © William H. Sadlier, Inc. All rights reserved.

# Finding Commission

Name ____________________

Date ____________________

Find the commission if the rate of commission is 4% and the amount of sale is $4280.

**Method 1**

$C = 4\%$ of $4280

$C = 0.04 \times \$4280$

$C = \$171.20$

**Method 2**

$\frac{C}{\$4280} = \frac{4}{100}$

$C \times 100 = \$4280 \times 4$

$C = \$17{,}120 \div 100 = \$171.20$

So the commission is $171.20.

**Find the commission.**

1. Amount sold = $950
   Rate of Commission = 3%
   Commission = ________

2. Amount sold = $1000
   Rate of Commission = 5.5%
   Commission = ________

3. Amount sold = $450
   Rate of Commission = 5%
   Commission = ________

4. Amount sold = $800
   Rate of Commission = 15%
   Commission = ________

5. Amount sold = $690
   Rate of Commission = 6.5%
   Commission = ________

6. Amount sold = $4500
   Rate of Commission = 2.75%
   Commission = ________

**PROBLEM SOLVING**

7. Mr. Bouchard sells television sets and VCRs at a $6\frac{1}{2}\%$ rate of commission. What is his commission on sales totaling $5245? ____________

8. Ms. Sumner sold $2578 worth of clothes last month. Her rate of commission is 5%. How much commission did she make? ____________

9. One week, Ms. Urbach sold $400 worth of reference books. If her commission was 12%, what did she receive for selling the books? ____________

10. Anita sold 8 recorders at $24.99 each and 12 records at $8.95 each. If her rate of commission was 6%, what was her total commission? ____________

 Copyright © William H. Sadlier, Inc. All rights reserved.

# Making Circle Graphs

Name ______________________

Date ______________________

**Fruit Sold**

144° — 8 Apples

216° — 12 Pears

**Percent of Total**

$\frac{8}{20} = \frac{n}{100}$

$8 \times 100 = 20 \times n$

$n = 800 \div 20 = 40$

$\frac{n}{100} = \frac{40}{100} = 40\%$

**Angle Measure**

40% of 360°

$40\% = \frac{40}{100} = \frac{2}{5}$

$\frac{2}{5} \times 360° = d$

$144° = d$

**Use the circle graph at the right to answer each.**

1. What percent of the budget is *not* food? ________

2. If the budget is based on $600 per week, how much money is set aside for recreation? for rent? ________

3. What percent of the budget is spent for food and rent? ________

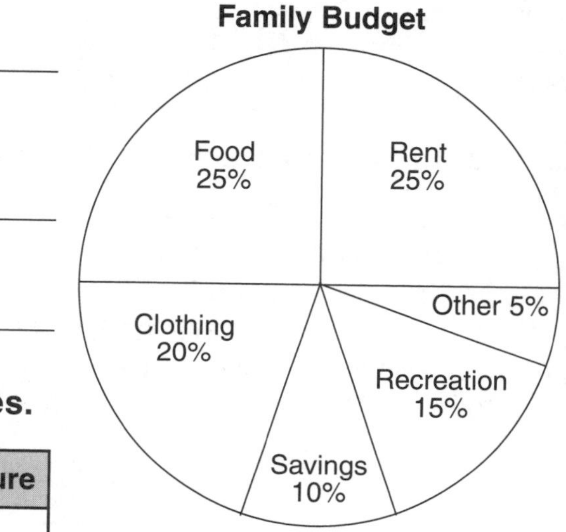


**Complete the chart showing Ron's vacation expenses.**

| | Item | Amount Spent | Percent of Total | Angle Measure |
|---|---|---|---|---|
| 4. | Hotel | $65.10 | | |
| 5. | Gas | $18.60 | | |
| 6. | Meals | $46.50 | | |
| 7. | Clothes | $37.20 | | |
| 8. | Other | $18.60 | | |

**Draw a circle graph showing the sale of fruit. Label your graph and include a title.**

9.

| Fruit | Number Sold |
|---|---|
| Plums | 150 |
| Apples | 210 |
| Pears | 30 |
| Peaches | 60 |
| Oranges | 150 |


Copyright © William H. Sadlier, Inc. All rights reserved.

# Problem-Solving Strategy: Write an Equation

Name ____________________

Date ____________________

Tailor's Department Store is offering a 15% discount on all merchandise. The original price of a sweater is \$35.40. How much is the discount? What is the sale price?

Let $D$ represent the discount.

$D$ = rate of discount × list price

$D = 15\% \times \$35.40$

$D = 0.15 \times \$35.40$

$D = \$5.31$

The discount is \$5.31.

Let $SP$ represent the sale price.

$SP$ = list price − discount

$SP = \$35.40 - \$5.31$

$SP = \$30.09$

The sale price is \$30.09.

**PROBLEM SOLVING. Write an equation. Do your work on a separate sheet of paper.**

1. In a survey, 480 people were asked if they were in favor of developing the town park. The results showed that 65% of the people surveyed were in favor. How many people were in favor of developing the town park?

   ____________________

2. Ms. Juliano sold 4 computer packages for the following amounts: \$1525, \$1250, \$2050, and \$3075. If her rate of commission on these 4 sales was 3%, what was her total commission for all 4 sales?

   ____________________

3. Of the 1566 people at the ball game, $66\frac{2}{3}\%$ sat on the home-team side. How many people sat on the visiting-team side?

   ____________________

4. Anita bought a book that cost \$14.95 plus 8% sales tax. Find the sales tax and total cost of the book.

   ____________________

5. There are 40 animals in the Perky Pet Store. Of these animals, 30% are dogs. How many of the animals are dogs?

   ____________________

6. A camera that costs \$250 is on sale for \$220. What is the percent of discount on the camera?

   ____________________

7. Blue Ribbon Supplies had a 20% sale on all horse items. Karen bought a hay net that had a list price of \$6.80. How much was the discount? What was the sale price?

   ____________________

8. The house that Shawna lives in is on a 15,000 square foot lot. If the house occupies about 8% of the lot, how many square feet does the house occupy?

   ____________________


Copyright © William H. Sadlier, Inc. All rights reserved.

# Integers

Name ______________________

Date ______________________

negative integers — zero — positive integers

$^{-}5$ $^{-}4$ $^{-}3$ $^{-}2$ $^{-}1$ 0 $^{+}1$ $^{+}2$ $^{+}3$ $^{+}4$ $^{+}5$

$^{-}4 < ^{-}2$ $^{-}1 < ^{+}1$ $^{+}2 < ^{+}4$

opposites

**Write the opposite of each integer.**

**1.** $^{-}6$ ____ **2.** $^{-}4$ ____ **3.** $^{+}5$ ____ **4.** $^{-}9$ ____ **5.** $^{+}12$ ____

**6.** $^{+}8$ ____ **7.** $^{+}2$ ____ **8.** $^{-}3$ ____ **9.** $^{-}15$ ____ **10.** $^{+}6$ ____

**11.** $^{-}24$ ____ **12.** $^{+}33$ ____ **13.** $^{-}17$ ____ **14.** $^{+}99$ ____ **15.** $^{-}7$ ____

**Write the integer that matches the letter on this number line.**

| A | B | C | D | E | F | G | H | I | J | K | L | M | N | O |
|---|---|---|---|---|---|---|---|---|---|---|---|---|---|---|
| $^{-}7$ | $^{-}6$ | $^{-}5$ | $^{-}4$ | $^{-}3$ | $^{-}2$ | $^{-}1$ | 0 | $^{+}1$ | $^{+}2$ | $^{+}3$ | $^{+}4$ | $^{+}5$ | $^{+}6$ | $^{+}7$ |

**16.** B ____ **17.** F ____ **18.** J ____ **19.** N ____ **20.** E ____

**21.** A ____ **22.** L ____ **23.** H ____ **24.** K ____ **25.** D ____

**Write the integer that is just before and just after the given integer.**

**26.** $^{+}10$ ________ **27.** $^{-}8$ ________ **28.** $^{-}1$ ________ **29.** $^{+}6$ ________

**30.** $^{-}5$ ________ **31.** $^{+}1$ ________ **32.** $^{+}12$ ________ **33.** $^{-}25$ ________

**34.** 0 ________ **35.** $^{-}16$ ________ **36.** $^{-}6$ ________ **37.** $^{+}5$ ________

**Write an integer to represent each situation.**

**38.** 52 feet below sea level? ____ **39.** deposit of $25 ____

**40.** 15-yard penalty? ____ **41.** 10° above zero ____

**42.** 8 feet above sea level? ____ **43.** withdrawal of $18 ____

**For exercise 41, describe an opposite situation. Write an integer to represent it.**

**44.** ______________________

**Use with Lesson 13-1, text pages 442–443.**

Copyright © William H. Sadlier, Inc. All rights reserved.

# Comparing and Ordering Integers

Name ____________________

Date ____________________

On a horizontal number line, any number is greater than a number to its left.

$^{-}5$ $^{-}4$ $^{-}3$ $^{-}2$ $^{-}1$ $0$ $^{+}1$ $^{+}2$ $^{+}3$ $^{+}4$ $^{+}5$

$^{-}4 < ^{-}2$ $\quad$ $^{-}1 < ^{+}1$ $\quad$ $^{+}2 < ^{+}4$

**Circle the greater integer.**

**1.** $^{+}8$, $^{+}6$ **2.** $^{-}1$, $^{-}4$ **3.** $^{-}3$, $^{+}2$ **4.** $^{+}5$, $^{-}5$

**5.** $^{-}6$, $^{-}7$ **6.** $^{-}3$, $0$ **7.** $^{+}1$, $^{-}4$ **8.** $0$, $^{+}4$

**9.** $^{+}3$, $^{-}8$ **10.** $^{-}6$, $^{-}2$ **11.** $^{+}7$, $^{+}10$ **12.** $^{+}4$, $^{-}8$

**Compare. Write <, =, or >.**

**13.** $^{+}6$ ____ $^{+}10$ **14.** $^{-}6$ ____ $^{-}3$ **15.** $^{+}10$ ____ $^{+}10$ **16.** $^{-}3$ ____ $0$

**17.** $^{+}5$ ____ $0$ **18.** $^{-}7$ ____ $^{+}7$ **19.** $^{+}4$ ____ $^{-}4$ **20.** $^{+}3$ ____ $^{-}1$

**21.** $^{+}7$ ____ $^{-}4$ **22.** $^{+}9$ ____ $0$ **23.** $^{-}5$ ____ $^{+}2$ **24.** $0$ ____ $0$

**25.** $^{+}2$ ____ $^{+}8$ **26.** $^{-}5$ ____ $^{-}1$ **27.** $^{-}6$ ____ $^{-}3$ **28.** $^{+}4$ ____ $0$

**Write in order from least to greatest.**

**29.** $0, ^{-}2, ^{+}4$ ____________________

**30.** $^{-}4, ^{+}6, ^{+}5$ ____________________

**31.** $^{+}10, ^{-}2, ^{+}6$ ____________________

**32.** $^{-}1, ^{-}9, ^{-}8, ^{+}2$ ____________________

**33.** $^{-}7, 0, ^{-}3, ^{+}2$ ____________________

**34.** $^{+}4, ^{-}1, ^{-}6$ ____________________

**35.** $^{-}2, ^{-}8, ^{-}6, ^{-}3$ ____________________

**36.** $^{+}7, ^{+}5, ^{+}1, ^{+}3$ ____________________

**Write in order from greatest to least.**

**37.** $^{-}7, ^{+}4, ^{-}5$ ____________________

**38.** $^{-}9, ^{-}10, ^{-}4$ ____________________

**39.** $0, ^{-}6, ^{-}11$ ____________________

**40.** $^{+}3, ^{+}5, ^{+}9$ ____________________

**41.** $^{+}2, ^{-}5, ^{+}8, ^{-}1$ ____________________

**42.** $^{-}3, ^{+}8, 0, ^{-}1$ ____________________

**43.** $0, ^{-}17, ^{+}2, ^{-}5$ ____________________

**44.** $^{+}1, ^{-}8, ^{+}9, ^{-}6$ ____________________

 Copyright © William H. Sadlier, Inc. All rights reserved.

# Addition Model for Integers

Name ______________________

Date ______________________

| **Add:** $^{+}4 + ^{+}1$ | **Add:** $^{-}2 + ^{-}4$ |
|---|---|
| $^{+}4$ [+][+][+][+] | $^{-}2$ [−][−] |
| $^{+}1$ [+] | $^{-}4$ [−][−][−][−] |
| Count the positives: 5. | Count the negatives: 6. |
| So $^{+}4 + ^{+}1 = ^{+}5$. | So $^{-}2 + ^{-}4 = ^{-}6$. |

**Write an addition sentence for each.**

**1.** $^{+}1$ [+]
$^{+}3$ [+][+][+]

______________

**2.** $^{-}6$ [−][−][−][−][−][−]
$^{-}2$ [−][−]

______________

**3.** $^{+}2$ [+][+]
$^{+}5$ [+][+][+][+][+]

______________

**4.** $^{-}4$ [−][−][−][−]
$^{-}3$ [−][−][−]

______________

**5.** $^{+}4$ [+][+][+][+]
$^{+}4$ [+][+][+][+]

______________

**6.** $^{-}1$ [−]
$^{-}2$ [−][−]

______________

**Add. You may draw positive and negative cards.**

**7.** $^{+}7 + ^{+}2 =$ _____

**8.** $^{-}5 + ^{-}4 =$ _____

**9.** $^{+}6 + ^{+}4 =$ _____

**10.** $^{-}5 + ^{-}1 =$ _____

**11.** $^{+}10 + ^{+}2 =$ _____

**12.** $^{-}8 + ^{-}3 =$ _____

**13.** $^{+}1 + ^{+}2 + ^{+}3 =$ _____

**14.** $^{-}4 + ^{-}3 + ^{-}4 =$ _____

**15.** $^{+}2 + ^{+}5 + ^{+}5 =$ _____

**Complete.**

**16.** The sum of two or more negative integers is a ______________________.

**17.** The sum of two or more positive integers is a ______________________.


Copyright © William H. Sadlier, Inc. All rights reserved.

# Another Addition Model

Name ______________________

Date ______________________

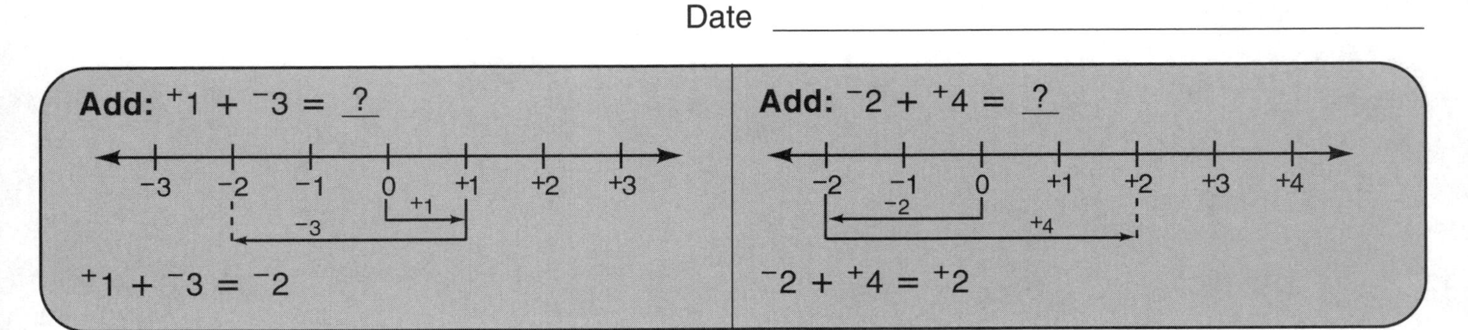


**Use a number line or integer cards to add.**

1. $^{+}3 + {}^{+}9 =$ ______
2. $^{+}6 + {}^{-}9 =$ ______
3. $^{-}10 + {}^{-}2 =$ ______
4. $^{-}6 + {}^{-}4 =$ ______
5. $^{+}2 + {}^{+}7 =$ ______
6. $^{-}13 + {}^{+}8 =$ ______
7. $^{+}11 + {}^{-}9 =$ ______
8. $^{+}7 + {}^{-}5 =$ ______
9. $^{-}6 + {}^{+}3 =$ ______
10. $^{+}8 + {}^{+}3 =$ ______
11. $^{-}3 + {}^{-}8 =$ ______
12. $^{-}7 + {}^{+}17 =$ ______
13. $^{+}4 + {}^{-}16 =$ ______
14. $^{-}8 + {}^{+}4 =$ ______
15. $^{-}4 + {}^{+}8 =$ ______
16. $^{+}1 + {}^{+}3 + {}^{-}4 =$ ______
17. $^{-}3 + {}^{+}2 + {}^{+}4 =$ ______
18. $^{-}3 + {}^{-}3 + {}^{-}3 =$ ______

**Write a number sentence for each number line.**

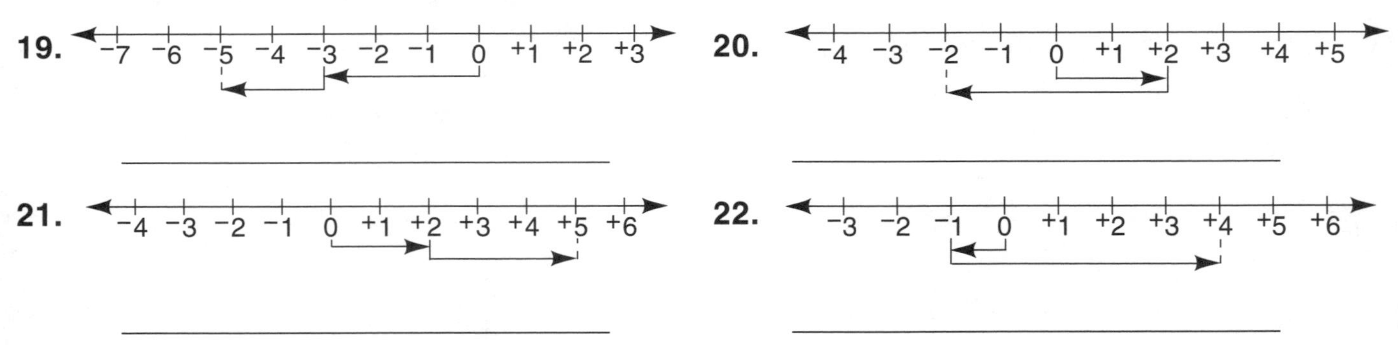


19. ______________________
20. ______________________
21. ______________________
22. ______________________

**PROBLEM SOLVING**

23. A quarterback gained 7 yards on one play. Then he lost 15 yards on the next play. What was his total yardage on the two plays? ______________________

24. In the first round of a game, Josie lost 8 points. In the second round, she earned 3 points. What was her score after the second round? ______________________

25. Curtis withdrew $12 from his savings account. The next week he deposited $16. Did he have more or less money in his account after he made the deposit than before he made the withdrawal? How much more or less? ______________________

 **Use with Lesson 13-4, text pages 448–449.** Copyright © William H. Sadlier, Inc. All rights reserved.

# Adding Integers

Name ____________________

Date ____________________

| **Like Signs** | **Unlike Signs** |
|---|---|
| $^{+}2 + {}^{+}4 = \underline{?} \rightarrow {}^{+}2 + {}^{+}4 = {}^{+}6$ | $^{-}8 + {}^{+}6 = \underline{?} \rightarrow 8 - 6 = 2 \rightarrow {}^{-}8 + {}^{+}6 = {}^{-}2$ |
| $^{-}2 + {}^{-}4 = \underline{?} \rightarrow {}^{-}2 + {}^{-}4 = {}^{-}6$ | $^{-}3 + {}^{+}9 = \underline{?} \rightarrow 9 - 3 = 6 \rightarrow {}^{-}3 + {}^{+}9 = {}^{+}6$ |

**Add.**

**1.** $^{+}6 + {}^{+}6 =$ _____
**2.** $^{+}3 + {}^{+}5 =$ _____
**3.** $^{-}4 + {}^{-}5 =$ _____
**4.** $^{-}1 + {}^{-}1 =$ _____
**5.** $^{-}3 + {}^{-}6 =$ _____
**6.** $^{+}9 + {}^{+}9 =$ _____
**7.** $^{-}20 + {}^{-}42 =$ _____
**8.** $^{+}17 + {}^{+}5 =$ _____
**9.** $^{+}11 + {}^{+}12 =$ _____
**10.** $^{-}5 + {}^{+}7 =$ _____
**11.** $^{-}4 + {}^{+}3 =$ _____
**12.** $^{+}5 + {}^{-}5 =$ _____
**13.** $^{+}1 + {}^{-}11 =$ _____
**14.** $^{-}8 + {}^{+}4 =$ _____
**15.** $^{+}13 + {}^{-}6 =$ _____
**16.** $^{-}4 + {}^{+}6 =$ _____
**17.** $^{-}15 + {}^{+}16 =$ _____
**18.** $^{-}23 + {}^{+}7 =$ _____
**19.** $(^{-}5 + {}^{-}6) + {}^{+}4 =$ _____
**20.** $^{+}4 + (^{+}3 + {}^{+}7) =$ _____
**21.** $(^{-}5 + {}^{-}7) + {}^{-}8 =$ _____

**Complete each table.**

| | Integer | Add $^{+}5$ |
|---|---|---|
| **22.** | $^{+}3$ | |
| **23.** | $^{-}9$ | |
| **24.** | $^{+}7$ | |
| **25.** | $^{-}4$ | |
| **26.** | $^{-}10$ | |

| | Integer | Add $^{-}4$ |
|---|---|---|
| **27.** | $^{+}5$ | |
| **28.** | $^{-}7$ | |
| **29.** | 0 | |
| **30.** | $^{-}6$ | |
| **31.** | $^{+}9$ | |

## PROBLEM SOLVING

**32.** A submarine was 10 ft below sea level. It went down 20 more feet. Write its new depth as an integer. ____________________

**33.** In April, Derek gained 2 pounds. In May, he lost 5 pounds. Write his total gain or loss in April and May as an integer. ____________________

**34.** The West Town football team gained 18 yd on one play and lost 5 yd on the next play. Write their net gain or loss for the two plays as an integer. ____________________

**35.** An anchor hung against the side of a boat 5 ft above sea level. A sailor lowered the anchor 22 ft. Write its depth as an integer. ____________________


Copyright © William H. Sadlier, Inc. All rights reserved.

# Subtracting Integers

Name ____________________

Date ____________________

| Subtract: $^{+}9 - ^{+}4$ | Subtract: $^{-}7 - ^{+}2$ | Subtract: $^{-}8 - ^{-}5$ |
|---|---|---|
| $^{+}9 - ^{+}4 = ^{+}9 + ^{-}4$ | $^{-}7 - ^{+}2 = ^{-}7 + ^{-}2$ | $^{-}8 - ^{-}5 = ^{-}8 + ^{+}5$ |
| $^{+}9 - ^{+}4 = ^{+}5$ | $^{-}7 - ^{+}2 = ^{-}9$ | $^{-}8 - ^{-}5 = ^{-}3$ |

**Write the appropriate addition sentence. Then add.**

**1.** $^{+}10 - ^{+}5 =$ ______________ = ______

**2.** $^{-}4 - ^{-}4 =$ ______________ = ______

**3.** $^{-}9 - ^{-}15 =$ ______________ = ______

**4.** $^{+}7 - ^{-}8 =$ ______________ = ______

**5.** $^{+}7 - ^{+}10 =$ ______________ = ______

**6.** $^{-}10 - ^{-}12 =$ ______________ = ______

**7.** $^{+}12 - ^{-}16 =$ ______________ = ______

**8.** $^{-}13 - ^{+}11 =$ ______________ = ______

**9.** $^{-}5 - ^{-}4 =$ ______________ = ______

**10.** $^{+}8 - ^{-}12 =$ ______________ = ______

**Subtract.**

**11.** $^{-}5 - ^{-}10 =$ ______

**12.** $^{-}6 - ^{-}8 =$ ______

**13.** $^{+}6 - ^{+}4 =$ ______

**14.** $^{+}7 - ^{-}17 =$ ______

**15.** $^{+}4 - ^{+}9 =$ ______

**16.** $^{-}8 - ^{-}7 =$ ______

**17.** $^{+}5 - ^{-}8 =$ ______

**18.** $^{-}9 - ^{-}16 =$ ______

**19.** $^{+}10 - ^{-}10 =$ ______

**20.** $^{+}17 - ^{+}15 =$ ______

**21.** $^{-}6 - ^{+}56 =$ ______

**22.** $^{-}1 - ^{-}5 =$ ______

**Circle the letter of the correct answer.**

**23.** $^{+}6 - ^{-}10$ **a.** $^{-}4$ **b.** $^{+}16$ **c.** $^{-}16$ **d.** $^{+}4$

**24.** $^{-}2 - ^{-}5$ **a.** $^{-}7$ **b.** $^{-}3$ **c.** $^{+}7$ **d.** $^{+}3$

**25.** $^{-}3 - ^{-}3$ **a.** 0 **b.** $^{-}6$ **c.** $^{+}6$ **d.** $^{+}9$

**26.** $^{+}11 - ^{-}3$ **a.** $^{+}14$ **b.** $^{+}8$ **c.** $^{-}14$ **d.** $^{-}8$

**27.** $^{+}17 - ^{+}13$ **a.** $^{+}30$ **b.** $^{-}30$ **c.** $^{+}4$ **d.** $^{-}4$

## PROBLEM SOLVING

**28.** Dave was exploring coral 3 ft below sea level. Then he swam to an underwater cave with an entrance 15 ft below sea level. How many feet is the cave entrance from the coral? ____________________

**29.** Death Valley is 282 feet below sea level. The Caspian Sea is 92 feet below sea level. Write the difference of the two depths as an integer. ____________________

 Copyright © William H. Sadlier, Inc. All rights reserved.

# Temperature

Name ____________________

Date ____________________

The temperature was $^-6$°C. It rose 23 degrees. What was the new temperature?

$^-6 + ^+23 = ?$

$^-6 + ^+23 = ^+17$ or 17

The new temperature was 17°C.

The temperature was 5°F. It dropped to $^-10$°F. How many degrees did the temperature drop?

$^+5 - ^-10 = ?$

$^+5 - ^-10 = ^+5 + ^+10$

$= ^+15$ or 15

The temperature dropped 15°F.

50°
40° ← Normal body temperature
30°
20° ← Average room temperature
10°
0° ← Water freezes
$^-10$°
$^-20$°
°C

**Circle the most reasonable temperature for each.**

1. an ice cube — **a.** 100°C **b.** 0°C **c.** 50°C **d.** 32°C
2. a spring day — **a.** 20°C **b.** 5°C **c.** 0°C **d.** 50°C
3. a cold drink — **a.** $^-10$°C **b.** 30°C **c.** 50°C **d.** 10°C
4. a cold winter day — **a.** 10°C **b.** 30°C **c.** 20°C **d.** $^-10$°C

**Compute the new temperature in each exercise.**

5. 23°C; falls 5° ______
6. 14°C; rises 8° ______
7. $^-10$°F; rises 6° ______
8. 0°F; drops 15° ______
9. $^-10$°C; decreases 5° ______
10. $^-7$°C; climbs 12° ______
11. $^-10$°C; increases 14° ______
12. $^-18$°C; rises 15° ______
13. 25°F; drops 29° ______

**PROBLEM SOLVING**

14. At 8:00 A.M. the temperature was 6°C. At noon it was $^-12$°C. How many degrees did the temperature drop? ______
15. What is the difference between normal body temperature, 37°C and a cold winter day, $^-10$°C? ______
16. At 6:00 A.M. the temperature was $^-18$°F. By noon it was 25°F. How many degrees did the temperature rise? ______
17. The temperature was $^-22$°F. It rose 16 degrees. What was the new temperature? ______


Copyright © William H. Sadlier, Inc. All rights reserved.

# Ordered Pairs of Numbers

Name ______________________

Date ______________________

first number → $n$ ; second number → $n + 4$

| $n$ | $n + 4$ | Ordered Pair |
|---|---|---|
| 4 | $4 + 4 = 8$ | (4, 8) |
| 2 | $2 + 4 = 6$ | (2, 6) |
| 0 | $0 + 4 = 4$ | (0, 4) |

Rule: Add 4 to first number to get second number.

**Complete each table.**

| | $n$ | $n + 3$ | Ordered Pair |
|---|---|---|---|
| **1.** | 8 | | |
| **2.** | 4 | | |
| **3.** | 6 | | |
| **4.** | 12 | | |
| **5.** | 9 | | |

| | $n$ | $n - 5$ | Ordered Pair |
|---|---|---|---|
| **6.** | 7 | | |
| **7.** | 10 | | |
| **8.** | 15 | | |
| **9.** | 12 | | |
| **10.** | 6 | | |

**Complete each ordered pair for the given rule.**

**11.** $n - 2$ **a.** (9, _____) **b.** (6, _____) **c.** (4, _____)

**12.** $n + 10$ **a.** (0, _____) **b.** (4, _____) **c.** (10, _____)

**13.** $n \times 2$ **a.** (5, _____) **b.** (8, _____) **c.** (0, _____)

**14.** $n \div 3$ **a.** (27, _____) **b.** (12, _____) **c.** (18, _____)

**The rule for the ordered pairs is given by the formula $A = 3 \times w$. Complete the tables.**

| | ***W*** | 5 | 8 | 10 | 1 | 4 | 12 | 6 |
|---|---|---|---|---|---|---|---|---|
| **15.** | ***A*** | | 24 | | | | | |
| **16.** | ***(W,A)*** | (5, 15) | | | | | | |

| | ***W*** | 7 | 11 | 2 | 9 | 20 | 15 | 13 |
|---|---|---|---|---|---|---|---|---|
| **17.** | ***A*** | | | | | | | |
| **18.** | ***(W,A)*** | | | | | | | |


Copyright © William H. Sadlier, Inc. All rights reserved.

# Graphing Ordered Pairs of Integers

Name ____________________

Date ____________________

(4, 2) locates point *A*.

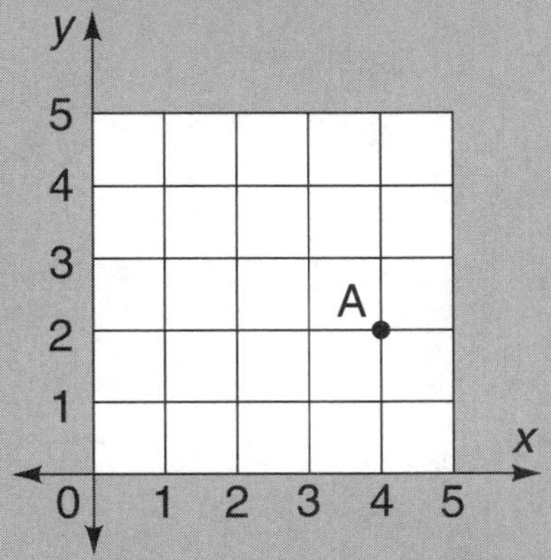

$(^{+}1, ^{-}3)$ locates point *B*.

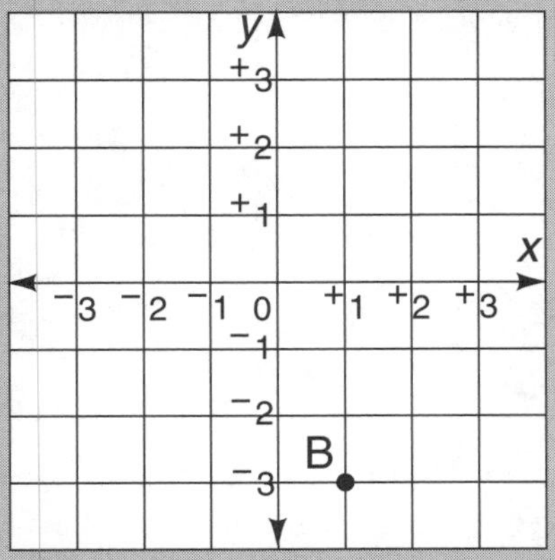

**Use the grid at the right for exercises 1–12. Name the ordered pair for each point.**

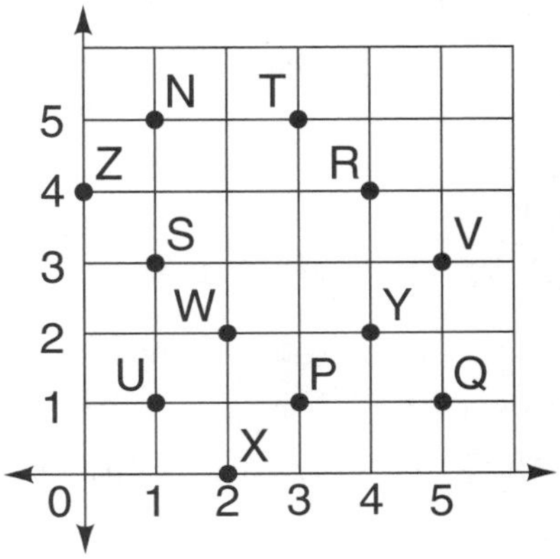

1. *Y* ________ 2. *Z* ________ 3. *U* ________

4. *R* ________ 5. *N* ________ 6. *V* ________

**Name the point for each ordered pair.**

7. (2, 2) ______ 8. (3, 1) ______ 9. (5, 1) ______

10. (1, 3) ______ 11. (2, 0) ______ 12. (3, 5) ______

**Use the grid at the right for exercises 13–20. Name the ordered pair for each point.**

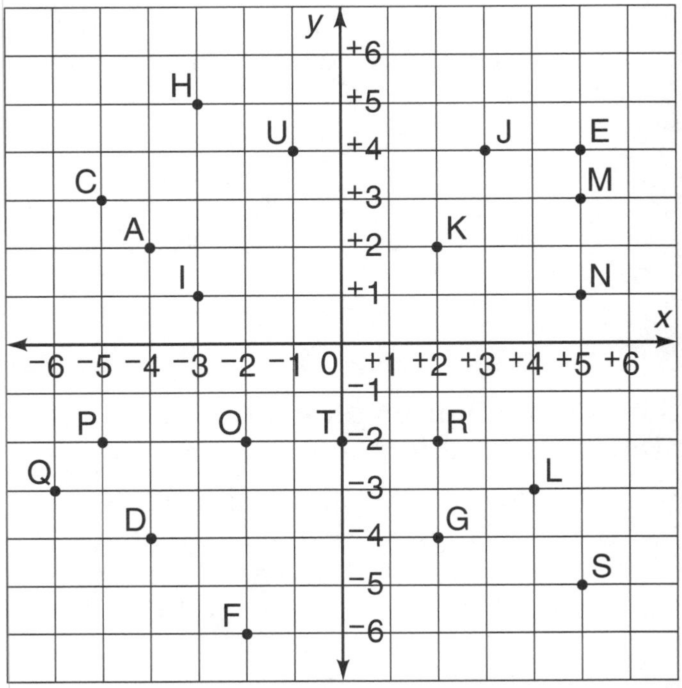

13. *A* ________ 14. *M* ________

15. *J* ________ 16. *Q* ________

17. *E* ________ 18. *H* ________

19. *F* ________ 20. *S* ________

21. *D* ________ 22. *P* ________

23. *K* ________ 24. *U* ________

**Name the point for each ordered pair.**

25. $(^{-}5, ^{+}3)$ ______ 26. $(^{-}2, ^{-}6)$ ______

27. $(0, ^{-}2)$ ______ 28. $(^{+}4, ^{-}3)$ ______

29. $(^{-}2, ^{-}2)$ ___ 30. $(^{+}2, ^{-}4)$ ___

 Copyright © William H. Sadlier, Inc. All rights reserved. 

# Graphing Transformations

Name ______________________________

Date ______________________________

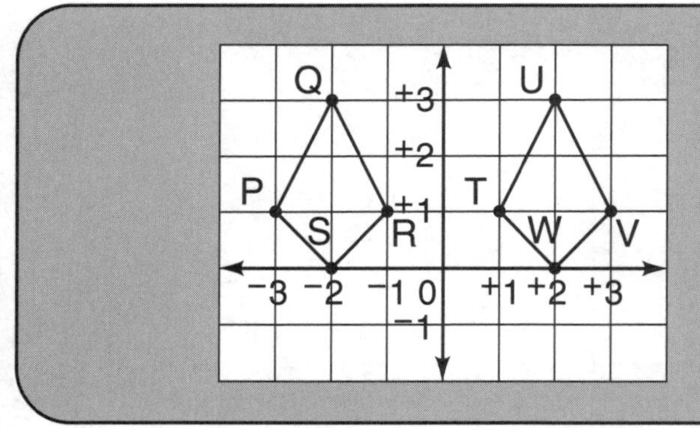


The coordinates of the vertices of the figure *PQRS* are $P(^-3, ^+1)$, $Q(^-2, ^+3)$, $R(^-1, ^+1)$, $S(^-2, 0)$.
Move right 4 units.

The coordinates of the vertices of the slide image are $T(^+1, ^+1)$, $U(^+2, ^+3)$, $V(^+3, ^+1)$, and $W(^+2, 0)$.

**Draw each triangle on the coordinate grid. Then draw the image and give the coordinates of its vertices.**

**1.** $D(^+4, ^+4)$, $E(^+4, ^+6)$, $F(^+6, ^+6)$
Move down 7 units.

______________________________

**2.** $A(0, ^-5)$, $B(^+3, ^-1)$, $C(^+4, ^-5)$
Move left 6 units.

______________________________

**3.** $P(^+1, ^+1)$, $Q(^+2, ^+5)$, $R(^+5, ^+1)$
Flip over y-axis line.

______________________________

**Use the grid below.**

**4.** Draw triangle $X(^+1, ^-4)$, $Y(0, 0)$, $Z(^+5, ^-4)$. Then draw the half-turn image of triangle *XYZ* rotated *clockwise* around the origin. Give the coordinates of its vertices.

______________________________

**5.** Draw the flip image over the *x*-axis of the image you drew in exercise 4. What did you discover?

______________________________

______________________________

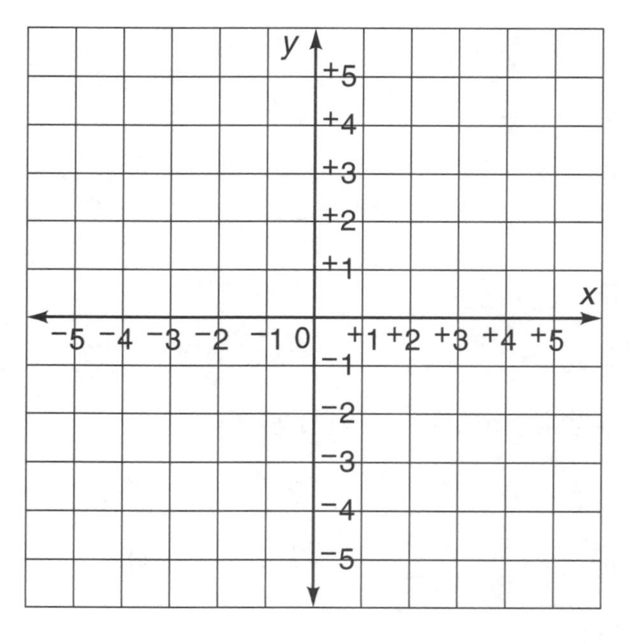


**Locate the figures on a coordinate grid. Tell whether the result is a slide or a flip.**

**6.** Join $A(^+1, ^-4)$, $B(0, 0)$, $C(^+5, ^-4)$. Make a new triangle *DEF* by using the same *x*-coordinates and adding $^+5$ to the *y*-coordinates. Join *D*, *E*, and *F*.

______________________________

 Copyright © William H. Sadler, Inc. All rights reserved.

# Problem-Solving Strategy: Make a Table

Name ____________________

Date ____________________

Anthony is making rose-and-iris bouquets. He is using 3 roses for every 4 irises. How many roses will be in a bouquet of 42 flowers?

Make a table.
Add to find the number of flowers in each bouquet.

| **Irises** | 4 | 8 | 12 | 16 | 20 | 24 |
|---|---|---|---|---|---|---|
| **Roses** | 3 | 6 | 9 | 12 | 15 | 18 |
| **Total Flowers** | 7 | 14 | 21 | 28 | 35 | 42 |

A bouquet of 42 flowers will have 18 roses.

**PROBLEM SOLVING. Do your work on a separate sheet of paper.**

1. Luanne inspected light bulbs. She found 3 out of every 95 light bulbs to be defective. How many defective light bulbs could she expect to find out of 760 light bulbs?

 ____________________

2. Theo recorded the temperature at 6:00 A.M. It was ⁻8°F. What would be the temperature at noon if it rose 7° every hour?

 ____________________

3. Steve conducted a survey and found that 48 out of 192 people bought Power Peanut Butter. How many people out of 1536 would you expect to buy Power Peanut Butter?

 ____________________

4. Tim is mixing blue and white paint to make a pale blue paint. He uses 2 quarts of blue for every 3 quarts of white paint. How many quarts of each color will he need to get 35 quarts of pale blue paint?

 ____________________

5. Every week Rochelle saves $12 in an envelope. Every other week she takes out $5 to spend on entertainment. How many weeks will it take her to have $76 in the envelope?

 ____________________

6. Sue Ellen is making a design using squares and circles. She uses 3 squares for every 5 circles. How many shapes in all will she use for her design if she uses 45 circles?

 ____________________

7. Mary had $150 in her savings account. She decided to add to it each week. The first week she added $5. For each week after that she added $3 more than the preceding week. How many weeks did it take her to save $335?

 ____________________

8. A recipe for one loaf of blueberry bread calls for 3 cups of flour, 2 cups of blueberries, and $\frac{1}{4}$ cup molasses. How much of each ingredient is needed for 10 loaves of blueberry bread?

 ____________________


Copyright © William H. Sadlier, Inc. All rights reserved.

# Problem Solving: Review of Strategies

Name ____________________

Date ____________________

**Solve. Do your work on a separate sheet of paper.**

1. Dale pasted 4 rows of 3 photographs on an album page. Each photograph is 5 cm long by 5 cm wide. He left 1.5 cm between each photograph and 2 cm as a border along each side of the page and 2 cm at the top and bottom of the page. What is the area of the page? What is the area of the page left uncovered when all the photographs are in place?

   ____________________

2. Dale has three polygons: a square, a rectangle, a parallelogram that is *not* a rectangle, a right triangle, an isoceles triangle, a trapezoid, an equilateral triangle, and a regular hexagon. If one of these polygons is selected at random, what is the probability that the polygon will have an odd number of sides? at least one acute angle? 3 or more sides? a straight angle?

   ____________________

3. On Monday Ralph began work at 8:15 A.M. He took a 15 minute break in the morning. He left for lunch at 12:55 P.M. and got back to work at 1:10 P.M. He took a 10 minute break in the afternoon. He left for home at 5:05 P.M. How long did Ralph work on Monday?

   ____________________

4. Ralph has three types of jackets: one flowered, one solid, one plaid. He wants to dress a model using one of the jackets and either a flared skirt or a straight skirt, slacks, or shorts. From how many different outfits does Ralph have to choose?

   ____________________

5. Lois has a collection of flags. She has 80 flags in all with 36 flags showing the color green, 52 flags showing red, 48 flags showing white and 20 flags showing black. What percent of flags in her collection shows the color green? red? white? black?

   ____________________

6. At 8:00 A.M. Lois noticed that the temperature was $^{-}14$°F. She recorded the temperature every 45 minutes after that. She discovered that the temperature rose 8°F every 45 minutes. The last temperature that Lois recorded was 26°F. What time was her last recording?

   ____________________

7. During a sale, a $600 piece of furniture is reduced by 10%. If the charge for packaging and delivery is 10% of the sale price, what is the final cost of this item?

   ____________________

8. Elena is 4 years younger than her brother, who is half her father's age. If her father is 5 years older than her mother, who is 47, how old is Elena?

   ____________________


Copyright © William H. Sadlier, Inc. All rights reserved.

# Algebraic Expressions

Name ______________________

Date ______________________

| English Expression | Mathematical Expression |
|---|---|
| 7 added to 10 | $10 + 7$ |
| 5 less than a number | $n - 5$ |
| 4 times a number | $4 \cdot x$ or $4x$ |
| 6 divided by a number | $6 \div a$ or $\frac{6}{a}$ |
| a number added to 6 | $6 + b$ |
| a number less than 5 | $5 - n$ |
| a number times four | $x \cdot 4$ |
| a number divided by 6 | $a \div 6$ or $\frac{a}{6}$ |

**Circle the letter of the correct mathematical expression for each English expression.**

1. 12 added to a number **a.** $12x$ **b.** $x + 12$ **c.** $x - 12$ **d.** $12 \div x$
2. a number divided by 2 **a.** $2n$ **b.** $n + 2$ **c.** $n \div 2$ **d.** $2 \div n$
3. 5 times a number **a.** $5x$ **b.** $x + 5$ **c.** $x - 5$ **d.** $5 + x$
4. 3 more than 7 **a.** $3 \cdot 7$ **b.** $7 \div 3$ **c.** $7 - 3$ **d.** $7 + 3$

**Translate each English expression to a mathematical expression. Use a variable when necessary.**

5. a number plus 17 ______
6. a number minus 14 ______
7. the product of $n$ and 5 ______
8. 1 less than a number ______
9. a number divided by 8 ______
10. the sum of a number and 44 ______
11. the quotient of 35 and $n$ ______
12. 4 less than $b$ ______
13. 36 divided by $z$ ______
14. 29 less than a number ______
15. 27 times a number ______
16. $n$ times 85 ______

**Translate to an algebraic expression. Use a variable when necessary.**

17. 4 less than twice a number ______
18. half a number plus 7 ______
19. double a number plus 20 ______
20. 3 times David's weight, plus 5 ______
21. the diameter divided by 2, minus 10 ______
22. 6 plus double the temperature ______
23. $\frac{1}{2}$ the width, minus 1 ______


Copyright © William H. Sadlier, Inc. All rights reserved.

# Equations

Name ____________________

Date ____________________

6 less than a number is 20.

$n - 6 = 20$

**Translate each to an equation. Use a variable when necessary.**

1. The product of $n$ and 16 is 32. ________
2. The sum of a number and 12 is 45. ________
3. $n$ plus 24 is 36. ________
4. 10 less than a number is 67. ________
5. A number tripled is 48. ________
6. The quotient of 100 divided by a number is 25. ________
7. A number times 14 is 70. ________
8. Two thirds of a number is 16. ________
9. The sum of 8 and a number is 40. ________
10. A number divided by 12 is 5. ________
11. The sum of $z$ and 16 is 26. ________
12. The product of 4 and $n$ is 48. ________

**Write an English sentence for each equation.**

13. $n \cdot 8 = 56$ ____________________
14. $x - 10 = 15$ ____________________
15. $35 = 7z$ ____________________
16. $100 + r = 125$ ____________________
17. $\frac{99}{p} = 3$ ____________________
18. $47 = 75 - b$ ____________________
19. $d \div 7 = 35$ ____________________

**Draw a balance-scale picture on a separate sheet of paper for each. Then solve.**

20. $c + 4 = 16$ ________
21. $12 = y - 1$ ________
22. $5 + g = 17$ ________


Copyright © William H. Sadlier, Inc. All rights reserved.

# Solving Equations: Guess and Test

Name ______________________

Date ______________________

To solve an equation, use the Guess and Test strategy to find the value of the variable that makes the equation true.

$x \div 4 = 26$

| Guess | Test | Conclusion | |
|---|---|---|---|
| 100 | $100 \div 4 = 25$ | $25 \neq 26$ | Try again. |
| 96 | $96 \div 4 = 24$ | $24 \neq 26$ | Try again. |
| 104 | $104 \div 4 = 26$ | $26 = 26$ | So $x = 104$. |

**Solve the equation. Use the Guess-and-Test strategy.**

1. $d + 5 = 13$ ______
2. $16 = c + 5$ ______
3. $4y = 20$ ______
4. $8 = 2m$ ______
5. $x + 18 = 25$ ______
6. $6s = 36$ ______
7. $3 + p = 15$ ______
8. $30 = 24 + k$ ______
9. $16 = 4d$ ______
10. $r \div 8 = 12$ ______
11. $x + 11 = 45$ ______
12. $14a = 56$ ______
13. $s - 8 = 15$ ______
14. $7m = 91$ ______
15. $z \div 5 = 7$ ______
16. $c + 34 = 98$ ______
17. $t \div 4 = 4$ ______
18. $48 = 3d$ ______

**Solve the equation. Watch the order of operations.**

19. $3a + 5 = 17$ ______
20. $4x - 1 = 19$ ______
21. $1 + 7c = 22$ ______
22. $8 = 2s - 12$ ______
23. $32 = 8 + 4h$ ______
24. $10y - 2 = 98$ ______
25. $1 + \frac{x}{2} = 4$ ______
26. $\frac{w}{4} - 6 = 0$ ______
27. $10 = \frac{s}{3} + 5$ ______

**Circle the letter of the correct equation. Then solve.**

28. Stanley bought 4 boxes of pens and 2 single pens. He had 50 pens in all. How many pens are in each box?

    **a.** $4 + 2p = 50$ **b.** $4p + 2 = 50$ **c.** $50 = (p \div 4) + 2$ ______________

29. Lucinda made 9 pennants. She put the same number of stars on each pennant. If she used 54 stars, how many stars did she put on each pennant?

    **a.** $54 = s \div 9$ **b.** $9 + s = 54$ **c.** $9s = 54$ ______________

30. Paul bought 18 tennis balls. There were 3 balls in each can. How many cans did he buy?

    **a.** $3t = 18$ **b.** $18 - 3 = t$ **c.** $3 + t = 18$ ______________


Copyright © William H. Sadlier, Inc. All rights reserved.

# Solving Equations: Add and Subtract

Name ______________________

Date ______________________

| Solve: $m + 5 = 12$ | Solve: $s - 3 = 8$ |
|---|---|
| Subtract 5 from both sides. | Add 3 to both sides. |
| $m + 5 - 5 = 12 - 5$ | $s - 3 + 3 = 8 + 3$ |
| $m = 7$ | $s = 11$ |

**Circle the letter of the operation to solve the equation. Then solve.**

**1.** $m - 3 = 10$ ____________

**a.** subtract 3 from both sides

**b.** add 3 to both sides

**c.** subtract 10 from both sides

**2.** $k + 6 = 15$ ____________

**a.** add 6 to both sides

**b.** add 15 to both sides

**c.** subtract 6 from both sides

**3.** $8 + y = 13$ ____________

**a.** add 13 to both sides

**b.** add 8 to both sides

**c.** subtract 8 from both sides

**4.** $s - 21 = 15$ ____________

**a.** add 21 to both sides

**b.** subtract 21 from both sides

**c.** add 15 to both sides

**Complete.**

**5.**

$23 = k + 12$

$23 - 12 = k + 12 -$ ______

______ $= k$

**6.**

$g - 42 = 13$

$g - 42 + 42 = 13 +$ ______

$g =$ ______

**7.**

$h - 7 = 17$

$h - 7 + 7 = 17 +$ ______

$h =$ ______

**8.**

$x + 12 = 34$

$x + 12 -$ ______ $= 34 -$ ______

$x =$ ______

**Solve and check.**

**9.** $7 + y = 25$ ______

**10.** $d - 6 = 33$ ______

**11.** $1.5 = s + 0.3$ ______

**12.** $p + 22 = 64$ ______

**13.** $12 = z - 10$ ______

**14.** $a - 15 = 0$ ______

**15.** $50 = b + 25$ ______

**16.** $p + 86 = 100$ ______

**17.** $x + .05 = 1$ ______

**18.** $f - 4.1 = 8.6$ ______

**19.** $h + \frac{3}{8} = \frac{5}{8}$ ______

**20.** $j - \frac{2}{10} = \frac{7}{10}$ ______

**21.** $k + 7 = 77$ ______

**22.** $m - 19 = 11$ ______

**23.** $r + .8 = 1.8$ ______

 **Use with Lesson 14-4, text pages 480–481.** Copyright © William H. Sadlier, Inc. All rights reserved.

# Solving Equations: Multiply and Divide

Name ______________________

Date ______________________

| Solve: $4n = 36$ | Solve: $x \div 5 = 3$ |
| --- | --- |
| Divide both sides by 4. | Multiply both sides by 5. |
| $\frac{4n}{4} = \frac{36}{4}$ | $x \div 5 \cdot 5 = 3 \cdot 5$ |
| $n = 9$ | $x = 15$ |

**Circle the letter of the operation to solve the equation. Then solve.**

1. $b \cdot 5 = 60$ ________
   a. multiply both sides by 5
   b. divide both sides by 5
   c. divide both sides by 60

2. $c \div 7 = 11$ ________
   a. multiply both sides by 11
   b. multiply both sides by 7
   c. divide both sides by 7

3. $3x = 18$ ________
   a. multiply both sides by 3
   b. divide both sides by 18
   c. divide both sides by 3

4. $z \div 4 = 8$ ________
   a. multiply both sides by 4
   b. multiply both sides by 8
   c. divide both sides by 4

**Complete.**

5. $d \div 3 = 16$
   $d \div 3 \times 3 = 16 \times$ ____
   $d =$ ____

6. $9f = 117$
   $\frac{9f}{9} = \frac{117}{\phantom{0}}$
   $f =$ ____

7. $4c = 56$
   $\frac{4c}{4} =$ ____
   $c =$ ____

8. $\frac{x}{2} = 24$
   $2 \cdot \frac{x}{2} =$ ____ $\cdot 24$
   $x =$ ____

**Solve and check.**

9. $r \div 6 = 10$ ______
10. $8 \cdot w = 120$ ______
11. $15 = \frac{e}{0.5}$ ______
12. $8m = 40$ ______
13. $x \div 6 = 6$ ______
14. $9 = t \div 8$ ______
15. $12 = \frac{1}{2}x$ ______
16. $\frac{b}{5} = 15$ ______
17. $3 = \frac{s}{0.3}$ ______


Copyright © William H. Sadlier, Inc. All rights reserved.

# Evaluating Formulas

Name ______________________________

Date ______________________________

Formula: a rule describing a mathematical relationship of two or more quantities.
To evaluate a formula, replace all variables except one with number values.

The distance formula is $d = r \times t$.
Evaluate $d = r \times t$ for $r$ (rate) when $d$ (distance) = 150 miles and $t$ (time) = $2\frac{1}{2}$ hours.

$d = r \times t$
$150 = r \times 2.5$ Write a related
$150 \div 2.5 = r$ division sentence.
$60 = r$

So the rate is 60 miles per hour.

The formula for velocity is $V = 32 \times t$ where $V$ is the velocity in feet per second that an object reaches after falling from a height for a certain time ($t$) in seconds.

A ball that is dropped from a building takes 6 seconds to hit the ground. What is the velocity of the ball?

$V = 32 \times t$
$V = 32 \times 6$
$V = 192$ feet per second

**Write the formula you would use. Then evaluate the formula to solve the problem.**

**1.** A bus travels at a rate of 55 miles per hour. How far does it travel in 8 hours?

______________________________

**2.** An egg falls from a tree and hits the ground in 3 seconds. What is its velocity?

______________________________

**3.** It takes a plane $5\frac{1}{2}$ hours to fly from New York to San Diego, a distance of 2805 miles. How fast does the plane fly?

______________________________

**4.** A hammer is dropped from a roof. Its velocity when it hits the ground is 256 feet per second. How long did it take the hammer to fall?

______________________________

**5.** A train makes a trip of 315 miles. If it travels at an average rate of 90 miles per hour, how long does the trip take?

______________________________

**6.** A pebble is dropped from a bridge. It takes 4 seconds for it to hit the surface of the water. What is its velocity?

______________________________

**Evaluate the formula to find the volume or height as indicated.**

**7.** $V = e \times e \times e$ or $V = e^3$ where $e = 25.5$ cm

$V =$ ______________

**8.** $V = l \times w \times h$ where $V = 5600$ ft$^3$, $l = 35$ ft, and $w = 8$ ft

$h =$ ______________

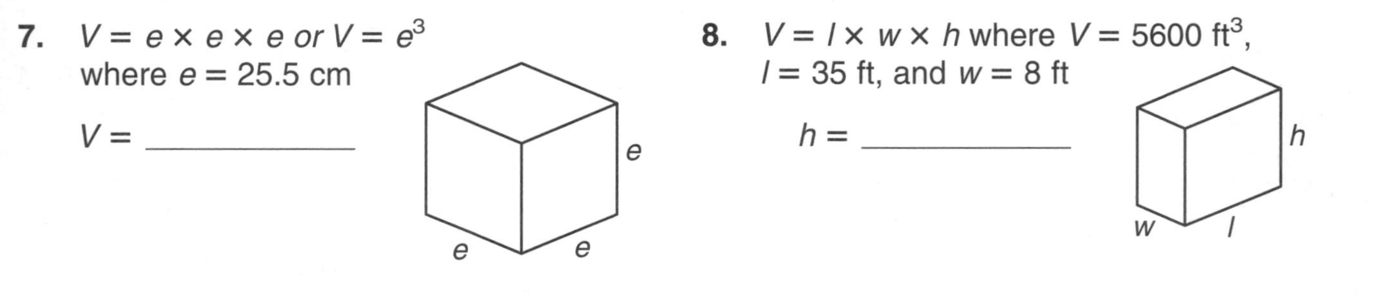


 Use with Lesson 14-6, text pages 484–485. Copyright © William H. Sadlier, Inc. All rights reserved.

# Evaluating Volume Formulas

Name ______________________

Date ______________________

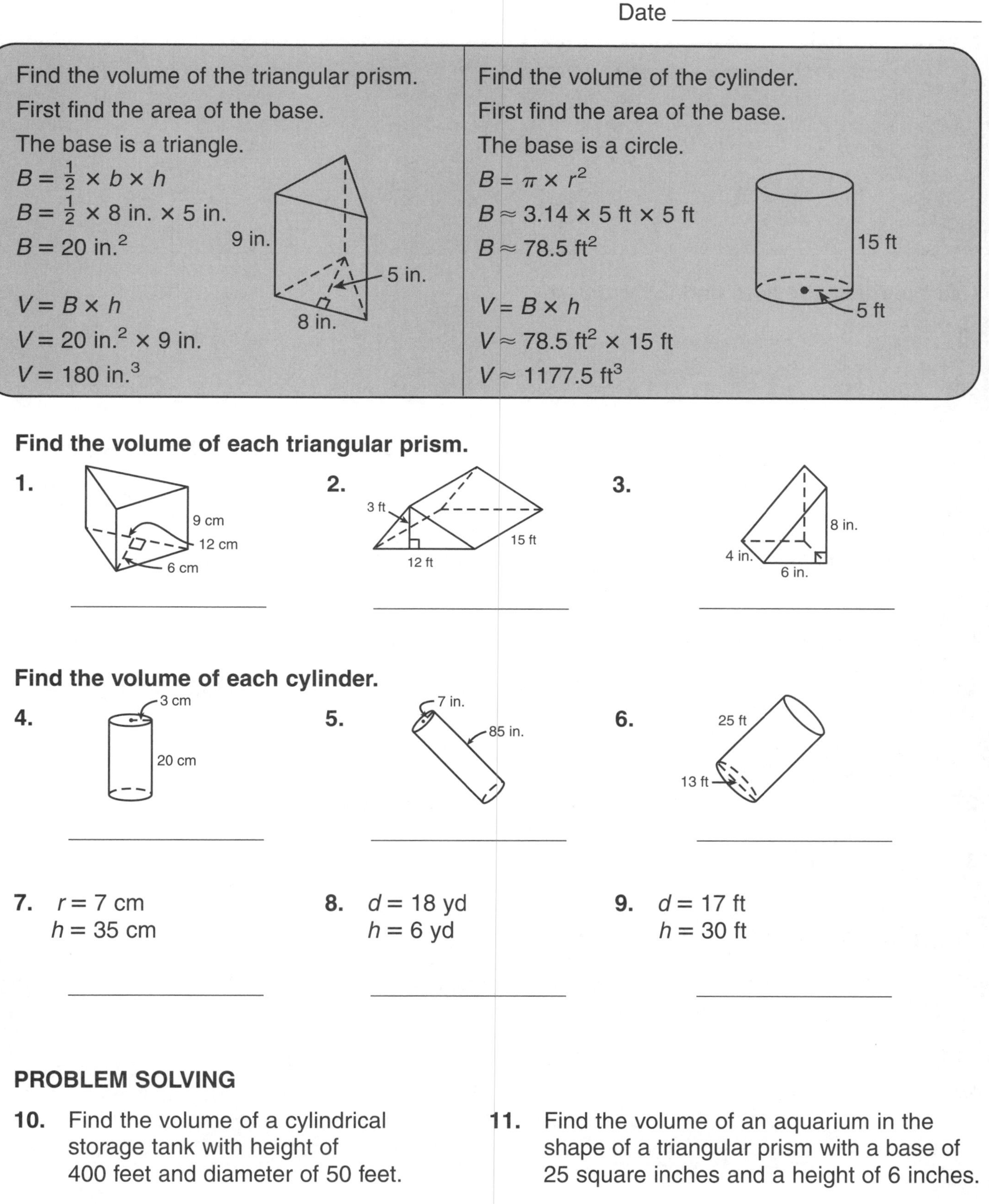


Find the volume of the triangular prism.
First find the area of the base.
The base is a triangle.

$B = \frac{1}{2} \times b \times h$

$B = \frac{1}{2} \times 8 \text{ in.} \times 5 \text{ in.}$

$B = 20 \text{ in.}^2$

$V = B \times h$

$V = 20 \text{ in.}^2 \times 9 \text{ in.}$

$V = 180 \text{ in.}^3$

Find the volume of the cylinder.
First find the area of the base.
The base is a circle.

$B = \pi \times r^2$

$B \approx 3.14 \times 5 \text{ ft} \times 5 \text{ ft}$

$B \approx 78.5 \text{ ft}^2$

$V = B \times h$

$V \approx 78.5 \text{ ft}^2 \times 15 \text{ ft}$

$V \approx 1177.5 \text{ ft}^3$

**Find the volume of each triangular prism.**

**1.** ______________

**2.** ______________

**3.** ______________

**Find the volume of each cylinder.**

**4.** ______________

**5.** ______________

**6.** ______________

**7.** $r = 7$ cm
$h = 35$ cm

______________

**8.** $d = 18$ yd
$h = 6$ yd

______________

**9.** $d = 17$ ft
$h = 30$ ft

______________

**PROBLEM SOLVING**

**10.** Find the volume of a cylindrical storage tank with height of 400 feet and diameter of 50 feet.

______________

**11.** Find the volume of an aquarium in the shape of a triangular prism with a base of 25 square inches and a height of 6 inches.

______________


Copyright © William H. Sadlier, Inc. All rights reserved.

# Multiplying Integers

Name ______________________

Date ______________________

| The product of two integers is *positive* if they have the *same* sign. | The product of two integers is *negative* if they have *different* signs. | The product of two integers is *zero* if one or both is *zero.* |
|---|---|---|
| $^{+}4 \times {}^{+}2 = {}^{+}8$<br>$^{-}4 \times {}^{-}2 = {}^{+}8$ | $^{-}4 \times {}^{+}2 = {}^{-}8$<br>$^{+}4 \times {}^{-}2 = {}^{-}8$ | $0 \times {}^{-}4 = 0$<br>$0 \times {}^{+}4 = 0$<br>$0 \times 0 = 0$ |

**Use the rules above to find the product.**

**1.** $^{-}5 \times {}^{+}3$ ____ **2.** $^{-}6 \times 0$ ____ **3.** $^{+}11 \times {}^{-}3$ ____ **4.** $^{-}8 \times {}^{-}5$ ____

**5.** $^{-}5 \times {}^{+}5$ ____ **6.** $^{+}3 \times {}^{+}3$ ____ **7.** $^{-}5 \times {}^{-}5$ ____ **8.** $0 \times {}^{+}2$ ____

**9.** $^{-}8 \times {}^{-}3$ ____ **10.** $^{+}9 \times {}^{+}3$ ____ **11.** $^{-}3 \times {}^{+}5$ ____ **12.** $^{+}10 \times {}^{+}3$ ____

**13.** $^{-}16 \times {}^{+}9$ ____ **14.** $^{+}12 \times {}^{-}18$ ____ **15.** $^{+}5 \times {}^{+}24$ ____ **16.** $^{-}13 \times {}^{-}14$ ____

**17.** $^{+}11 \times {}^{-}12$ ____ **18.** $^{-}36 \times 0$ ____ **19.** $^{-}12 \times {}^{-}12$ ____ **20.** $0 \times {}^{+}48$ ____

**Use the rules above to insert the sign of the underlined factor for the given product.**

**21.** $^{-}2 \times {}^{\square}\underline{5} = {}^{-}10$ **22.** $^{+}9 \times {}^{\square}\underline{8} = {}^{+}72$ **23.** $^{+}8 \times {}^{\square}\underline{11} = {}^{-}88$

**24.** $^{\square}\underline{6} \times {}^{+}5 = {}^{-}30$ **25.** $^{\square}\underline{8} \times {}^{-}8 = {}^{+}64$ **26.** $^{\square}\underline{3} \times {}^{-}7 = {}^{-}21$

**27.** $^{+}13 \times {}^{\square}\underline{11} = {}^{+}143$ **28.** $^{\square}\underline{12} \times {}^{-}15 = {}^{-}180$ **29.** $^{+}8 \times {}^{\square}\underline{27} = {}^{-}216$

**30.** $^{\square}\underline{16} \times {}^{-}16 = {}^{-}256$ **31.** $^{-}7 \times {}^{\square}\underline{23} = {}^{-}161$ **32.** $^{\square}\underline{18} \times {}^{+}26 = {}^{-}468$

## PROBLEM SOLVING

**33.** From the bank account activity shown in the table, write an integer sentence for each transaction to get the given result.

| Transaction | Sentence | Result |
|---|---|---|
| 3 deposits of 75 | | $^{+}\$225$ |
| 2 withdrawals of $35 | | $^{-}\$70$ |

**34.** A computer stock drops 2 points each day for ten days. What is the change over the ten days, written as an integer?

______________________

**35.** A tank was pumping out gasoline at a rate of 35 gallons an hour. What was the output of gas over an eight hour period, written as an integer?

______________________

 **Use with Lesson 14-8, text pages 488–489.** Copyright © William H. Sadlier, Inc. All rights reserved.

# Dividing Integers

Name ______________________

Date ______________________

| The quotient of two integers is *positive* if the integers have the *same* sign. | The quotient of two integers is *negative* if the integers have *different* signs. | The quotient of two integers is *zero* if the dividend is *zero.* |
|---|---|---|
| $^{+}18 \div {}^{+}9 = {}^{+}2$<br>$^{-}18 \div {}^{-}9 = {}^{+}2$ | $^{-}14 \div {}^{+}2 = {}^{-}7$<br>$^{+}14 \div {}^{-}2 = {}^{-}7$ | $0 \div {}^{+}8 = 0$<br>$0 \div {}^{-}8 = 0$<br>$0 \div 0 =$ impossible |

**Use the rules above to find each quotient.**

**1.** $^{+}24 \div {}^{-}8$ ____ **2.** $^{-}20 \div {}^{-}2$ ____ **3.** $^{+}36 \div {}^{-}9$ ____ **4.** $^{-}32 \div {}^{-}4$ ____

**5.** $^{-}48 \div {}^{+}6$ ____ **6.** $0 \div {}^{-}1$ ____ **7.** $^{+}10 \div {}^{-}2$ ____ **8.** $^{+}99 \div 0$ ____

**9.** $\frac{^{-}16}{^{-}4}$ ____ **10.** $\frac{^{+}72}{^{-}9}$ ____ **11.** $\frac{0}{^{-}10}$ ____ **12.** $\frac{^{-}50}{^{-}5}$ ____

**13.** $\frac{^{-}64}{^{+}8}$ ____ **14.** $\frac{^{-}54}{^{-}9}$ ____ **15.** $\frac{^{-}1}{0}$ ____ **16.** $\frac{^{-}85}{^{+}5}$ ____

**Complete each chart. Write the rule.**

**17.**

| IN | OUT |
|---|---|
| $^{+}12$ | $^{+}4$ |
| $^{+}36$ | $^{+}12$ |
| $^{+}18$ | $^{+}6$ |
| $^{-}21$ | ___ |
| $^{-}30$ | ___ |
| ___ | $^{-}11$ |

Rule: IN ÷ __ = OUT

**18.**

| IN | OUT |
|---|---|
| $^{-}30$ | $^{+}6$ |
| $^{-}45$ | $^{+}9$ |
| $^{-}40$ | $^{+}8$ |
| $^{+}20$ | ___ |
| $^{+}25$ | ___ |
| ___ | $^{+}15$ |

Rule: IN ÷ __ = OUT

**Find the value of each expression if $a = {}^{+}6$, $b = {}^{-}3$, $c = 0$, and $d = {}^{-}2$. (*Hint*: Use the rules for order of operations.)**

**19.** $a + b \cdot d$ ___ **20.** $\frac{c}{d}$ ___ **21.** $b - c + \frac{a}{d}$ ___ **22.** $\frac{c}{d} - \frac{c}{b}$ ___

**23.** $a \div (c \cdot d)$ ____ **24.** $c - a$ ___ **25.** $2b + \frac{c}{a}$ ___ **26.** $(b \cdot d) \div a$ ___

**PROBLEM SOLVING**

**27.** A stock dropped 32 points in 8 days. If the stock dropped at the same rate each day, what is the rate per day?

______________________

**28.** A diver is at depth of 350 meters. If she ascends at a rate of 5 meters per minute, how long will it take her to reach the surface?

______________________


Copyright © William H. Sadlier, Inc. All rights reserved.

# Equations with Integers

Name ______________________________

Date ______________________________

To solve an addition or subtraction equation using the *Guess and Test* strategy:

Solve: $n - {}^{-}5 = {}^{+}12$

| Guess | Test | Conclusion |
|---|---|---|
| $^{+}17$ | $^{+}17 - {}^{-}5 = {}^{+}22$ | $^{+}22 > {}^{+}12$ |
| $^{-}7$ | $^{-}7 - {}^{-}5 = {}^{-}2$ | $^{-}2 < {}^{+}12$ |
| $^{+}7$ | $^{+}7 - {}^{-}5 = {}^{+}12$ | $^{+}12 = {}^{+}12$ |

So $n = {}^{+}7$.

To solve an addition or subtraction equation using replacement numbers:

Solve: $n + {}^{-}5 = {}^{-}36$

Replacement numbers for $n$: $^{-}41$, $^{+}41$, $^{-}31$

| | | |
|---|---|---|
| Try $^{-}41$: | $^{-}41 + {}^{-}5 = {}^{-}46$ | No |
| Try $^{+}41$: | $^{+}41 + {}^{-}5 = {}^{+}36$ | No |
| Try $^{-}31$: | $^{-}31 + {}^{-}5 = {}^{-}36$ | Yes |

So $n = {}^{-}31$.

To solve multiplication and division equations, use the *Guess and Test* strategy.

Solve: $^{-}5y = {}^{+}60$

| | | |
|---|---|---|
| Try $^{+}12$: | $^{-}5 \cdot {}^{+}12 = {}^{-}60$ | No |
| Try $^{-}10$: | $^{-}5 \cdot {}^{-}10 = {}^{+}50$ | No |
| Try $^{-}12$: | $^{-}5 \cdot {}^{-}12 = {}^{+}60$ | Yes |

So $y = {}^{-}12$.

Solve: $t \div {}^{-}3 = {}^{-}6$

| | | |
|---|---|---|
| Try $^{+}9$: | $^{+}9 \div {}^{-}3 = {}^{-}3$ | No |
| Try $^{-}18$: | $^{-}18 \div {}^{-}3 = {}^{+}6$ | No |
| Try $^{+}18$: | $^{+}18 \div {}^{-}3 = {}^{-}6$ | Yes |

So $t = {}^{+}18$.

**Solve each equation. Use the *Guess and Test* strategy.**

**1.** $y + {}^{-}3 = {}^{-}13$ ______

**2.** $r + {}^{+}8 = {}^{+}40$ ______

**3.** $0 + x = {}^{-}35$ ______

**4.** $c + {}^{+}7 = {}^{-}45$ ______

**5.** $^{-}2 + d = {}^{+}15$ ______

**6.** $^{-}9 = n + {}^{+}20$ ______

**Solve each equation. Use these replacement numbers for *s*: $^{+}2$, $^{-}2$, 0, $^{+}6$, $^{-}6$. (Hint: Watch for − and +.)**

**7.** $s - {}^{+}6 = {}^{-}8$ ______

**8.** $s + {}^{+}8 = {}^{+}6$ ______

**9.** $s - {}^{-}8 = {}^{+}2$ ______

**10.** $s - {}^{-}6 = {}^{+}6$ ______

**11.** $^{+}2 = s + 0$ ______

**12.** $s - {}^{-}2 = {}^{+}8$ ______

**Solve each equation. Use the *Guess and Test* strategy.**

**13.** $^{+}5x = {}^{-}125$ ______

**14.** $u \div {}^{-}12 = {}^{+}12$ ______

**15.** $^{-}9y = 0$ ______

**16.** $m \div {}^{-}5 = {}^{-}16$ ______

**17.** $^{-}13a = {}^{+}39$ ______

**18.** $x \div {}^{+}9 = {}^{-}3$ ______

**PROBLEM SOLVING Write an equation and solve.**

**19.** Seven more than a number, $n$, equals $^{+}34$. What is the number?

______________________

**20.** A number $r$ times $^{-}10$ equals $^{+}130$. What is the number?

______________________


Copyright © William H. Sadlier, Inc. All rights reserved.

# Function Tables

Name ______________________________

Date ______________________________

For a linear function the graphs of the ordered pairs are points that form a straight line. To graph a function on a coordinate grid:

Make a function table for $y = {}^-3x + {}^-2$.

| $x$ | ${}^-3x + {}^-2$ | $y$ |
|---|---|---|
| ${}^-2$ | ${}^-3({}^-2) + {}^-2 = {}^+6 + {}^-2 = {}^+4$ | ${}^+4$ |
| ${}^-1$ | ${}^-3({}^-1) + {}^-2 = {}^+3 + {}^-2 = {}^+1$ | ${}^+1$ |
| 0 | ${}^-3(0) + {}^-2 = 0 + {}^-2 = {}^-2$ | ${}^-2$ |
| ${}^+1$ | ${}^-3({}^+1) + {}^-2 = {}^-3 + {}^-2 = {}^-5$ | ${}^-5$ |

Ordered pairs: $({}^-2, {}^+4)$, $({}^-1, {}^+1)$, $(0, {}^-2)$, $({}^+1, {}^-5)$

These ordered pairs are solutions of $y = {}^-3x + {}^-2$ because when you substitute for $x$ and $y$, you get a true statement.

Use the ordered pairs to graph $y = {}^-3x + {}^-2$ on a coordinate grid.

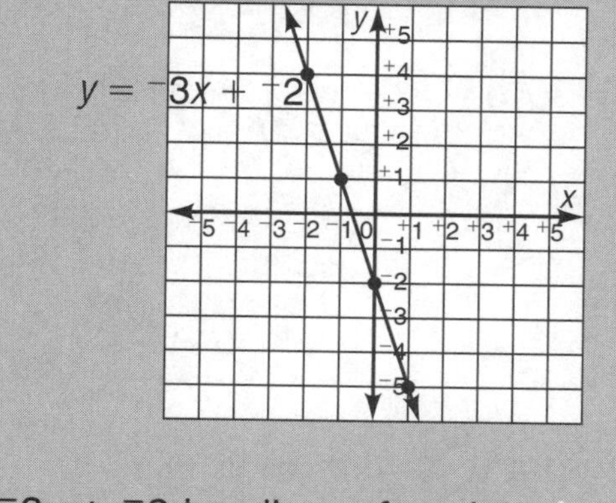


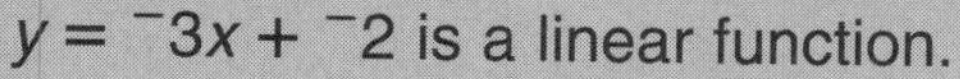
$y = {}^-3x + {}^-2$ is a linear function.

**1.** Complete the function table for $y = {}^-2x + {}^-4$. Then graph on a coordinate grid. Is the function a linear function?

| $x$ | ${}^-2x + {}^-4$ | $y$ |
|---|---|---|
| ${}^-2$ | | |
| ${}^-1$ | | |
| 0 | | |

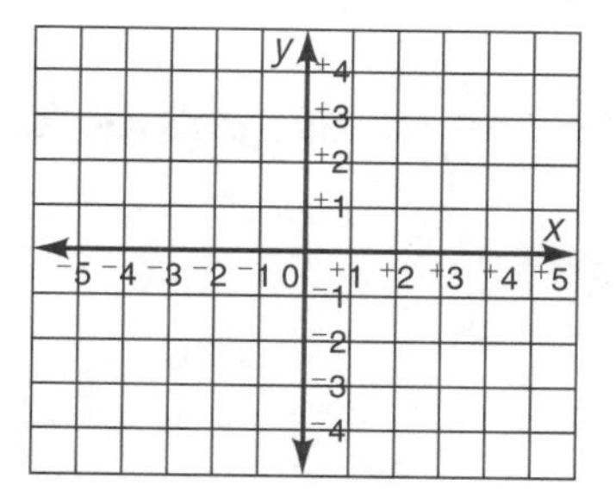

**Graph each function on a coordinate grid. Use ${}^+1$, 0, ${}^-1$, ${}^-2$ for $x$.**

**2.** $y = x$

**3.** $y = {}^-2x + {}^-1$

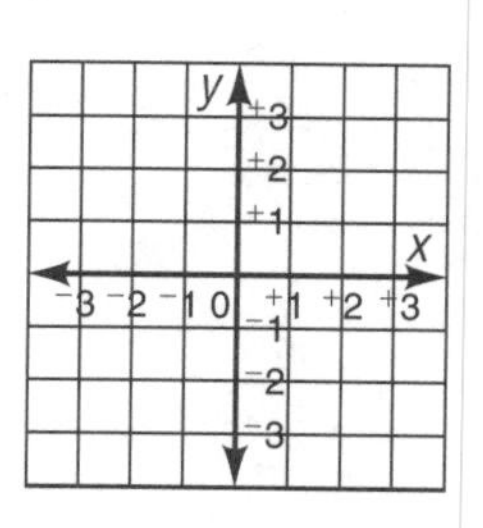

**4.** $y = x + {}^-1$

**PROBLEM SOLVING**

**5.** A storm is following a linear path. Twice the horizontal distance ($x$) plus 2 miles is the vertical distance ($y$). Write a linear function for the path of the storm.

______________________________

**6.** A car service charges a basic fee of $5 plus $8 per mile. Use $x$ for the number of miles. Write a linear function for finding the cost of a ride ($y$).

______________________________

 Copyright © William H. Sadlier, Inc. All rights reserved. 

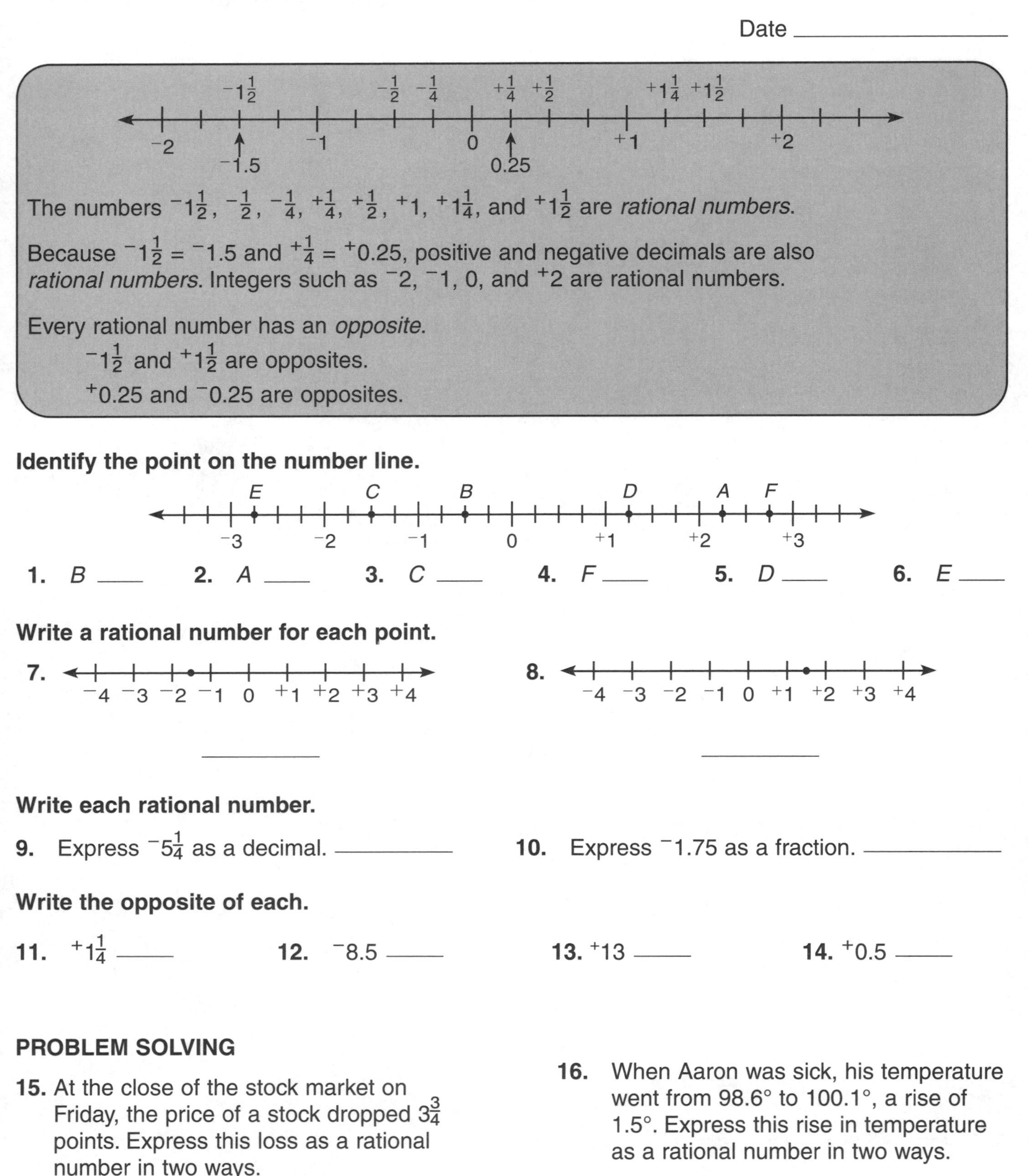

# Rational Numbers: Number Line

Name ____________________

Date ____________________



The numbers $^{-}1\frac{1}{2}$, $^{-}\frac{1}{2}$, $^{-}\frac{1}{4}$, $^{+}\frac{1}{4}$, $^{+}\frac{1}{2}$, $^{+}1$, $^{+}1\frac{1}{4}$, and $^{+}1\frac{1}{2}$ are *rational numbers.*

Because $^{-}1\frac{1}{2} = {}^{-}1.5$ and $^{+}\frac{1}{4} = {}^{+}0.25$, positive and negative decimals are also *rational numbers.* Integers such as $^{-}2$, $^{-}1$, 0, and $^{+}2$ are rational numbers.

Every rational number has an *opposite.*

$^{-}1\frac{1}{2}$ and $^{+}1\frac{1}{2}$ are opposites.

$^{+}0.25$ and $^{-}0.25$ are opposites.

**Identify the point on the number line.**



**1.** *B* ____ **2.** *A* ____ **3.** *C* ____ **4.** *F* ____ **5.** *D* ____ **6.** *E* ____

**Write a rational number for each point.**

**7.**



__________

**8.**



__________

**Write each rational number.**

**9.** Express $^{-}5\frac{1}{4}$ as a decimal. __________

**10.** Express $^{-}1.75$ as a fraction. __________

**Write the opposite of each.**

**11.** $^{+}1\frac{1}{4}$ ____ **12.** $^{-}8.5$ ____ **13.** $^{+}13$ ____ **14.** $^{+}0.5$ ____

**PROBLEM SOLVING**

**15.** At the close of the stock market on Friday, the price of a stock dropped $3\frac{3}{4}$ points. Express this loss as a rational number in two ways.

__________________

**16.** When Aaron was sick, his temperature went from 98.6° to 100.1°, a rise of 1.5°. Express this rise in temperature as a rational number in two ways.

__________________

 Copyright © William H. Sadlier, Inc. All rights reserved.

# Compare/Order Rational Numbers

Name ______________________

Date ______________________

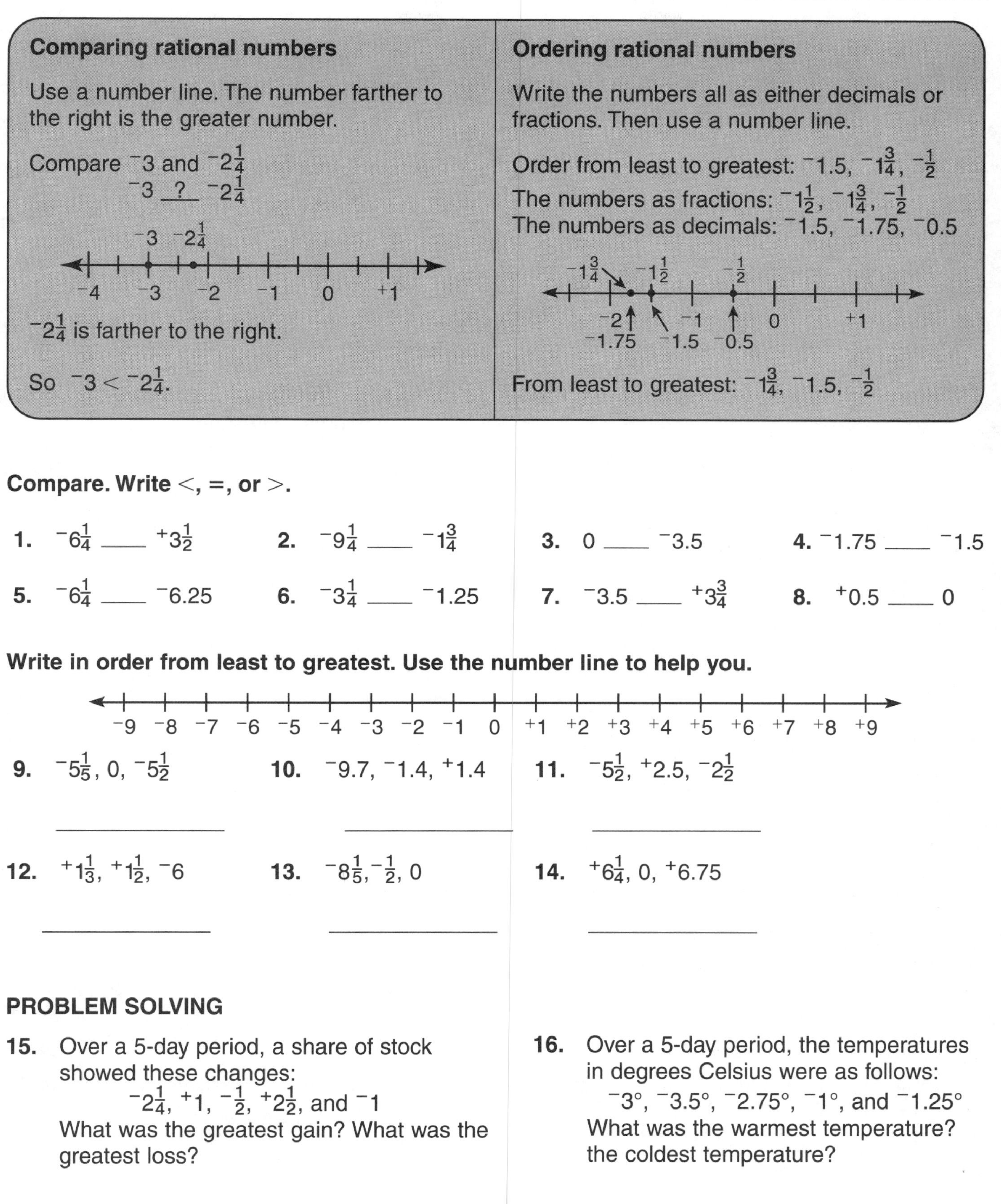

**Comparing rational numbers**

Use a number line. The number farther to the right is the greater number.

Compare $^{-}3$ and $^{-}2\frac{1}{4}$

$^{-}3$ __?__ $^{-}2\frac{1}{4}$



$^{-}2\frac{1}{4}$ is farther to the right.

So $^{-}3 < {}^{-}2\frac{1}{4}$.

**Ordering rational numbers**

Write the numbers all as either decimals or fractions. Then use a number line.

Order from least to greatest: $^{-}1.5$, $^{-}1\frac{3}{4}$, $^{-}\frac{1}{2}$

The numbers as fractions: $^{-}1\frac{1}{2}$, $^{-}1\frac{3}{4}$, $^{-}\frac{1}{2}$

The numbers as decimals: $^{-}1.5$, $^{-}1.75$, $^{-}0.5$



From least to greatest: $^{-}1\frac{3}{4}$, $^{-}1.5$, $^{-}\frac{1}{2}$

**Compare. Write <, =, or >.**

1. $^{-}6\frac{1}{4}$ ____ $^{+}3\frac{1}{2}$
2. $^{-}9\frac{1}{4}$ ____ $^{-}1\frac{3}{4}$
3. 0 ____ $^{-}3.5$
4. $^{-}1.75$ ____ $^{-}1.5$
5. $^{-}6\frac{1}{4}$ ____ $^{-}6.25$
6. $^{-}3\frac{1}{4}$ ____ $^{-}1.25$
7. $^{-}3.5$ ____ $^{+}3\frac{3}{4}$
8. $^{+}0.5$ ____ 0

**Write in order from least to greatest. Use the number line to help you.**



9. $^{-}5\frac{1}{5}$, 0, $^{-}5\frac{1}{2}$ ____________
10. $^{-}9.7$, $^{-}1.4$, $^{+}1.4$ ____________
11. $^{-}5\frac{1}{2}$, $^{+}2.5$, $^{-}2\frac{1}{2}$ ____________
12. $^{+}1\frac{1}{3}$, $^{+}1\frac{1}{2}$, $^{-}6$ ____________
13. $^{-}8\frac{1}{5}$, $^{-}\frac{1}{2}$, 0 ____________
14. $^{+}6\frac{1}{4}$, 0, $^{+}6.75$ ____________

**PROBLEM SOLVING**

15. Over a 5-day period, a share of stock showed these changes:
$^{-}2\frac{1}{4}$, $^{+}1$, $^{-}\frac{1}{2}$, $^{+}2\frac{1}{2}$, and $^{-}1$
What was the greatest gain? What was the greatest loss?
______________________

16. Over a 5-day period, the temperatures in degrees Celsius were as follows:
$^{-}3°$, $^{-}3.5°$, $^{-}2.75°$, $^{-}1°$, and $^{-}1.25°$
What was the warmest temperature? the coldest temperature?
______________________


Copyright © William H. Sadlier, Inc. All rights reserved.

# Problem-Solving Strategy: More Than One Solution

Name ____________________

Date ____________________

The area of a rectangle is 24 $ft^2$. How many possible whole-number dimensions result in an area of 24 $ft^2$?

Make a table. Use the formula $A = \ell \times w$.

So $24\ ft^2 = \ell \times w$.

| $\ell$ | 24 ft | 12 ft | 8 ft | 6 ft |
|---|---|---|---|---|
| $w$ | 1 ft | 2 ft | 3 ft | 4 ft |

There are four possible whole-number dimensions for a rectangle that result in an area of 24 $ft^2$.

**PROBLEM SOLVING Do your work on a separate sheet of paper.**

1. Roberto wants to have 12 shapes in a design using trapezoids and hexagons. If he uses each shape at least 3 times, how many different ways can he choose which shape to use?

   ____________________

2. Kyoko bought a dress for $36.11 plus 8% sales tax. How much did she pay for the dress? She gave the clerk a fifty dollar bill to pay for her purchase. How many different ways can she receive her change without using any coins?

   ____________________

3. What operations can you use to make the number sentence 2 _?_ 2 = 4 true?

   ____________________

4. The area of a rectangle is 42 $cm^2$. How many different sets of whole-number dimensions result in an area of 42 $cm^2$?

   ____________________

5. The sum of two single-digit integers is $^-6$. What are the possible addends that result in this sum?

   ____________________

6. Hassan tripled the sum of two negative integers. His answer was $^-24$. What are the possible addends?

   ____________________

7. Andrea bought 30 ft of fencing. How many different rectangles might she enclose, using whole-number dimensions?

   ____________________

8. A rectangular prism has a volume of 12 $cm^3$. How many different sets of whole-number dimensions result in this volume?

   ____________________

9. A triangle has an area of 30 $cm^2$. How many different sets of whole-number dimensions result in this area?

   ____________________

10. Dennis bought a book for $19.50 and paid with a twenty-dollar bill. In how many different ways might he have received change if he received no pennies?

    ____________________

 Copyright © William H. Sadlier, Inc. All rights reserved.